Non-Relativistic Quantum Dynamics

MATHEMATICAL PHYSICS STUDIES

A SUPPLEMENTARY SERIES TO
LETTERS IN MATHEMATICAL PHYSICS

Editors:

D. ARNAL, *Université de Dijon, France*
M. FLATO, *Université de Dijon, France*
M. GUENIN, *Institut de Physique Théorique, Geneva, Switzerland*
R. RĄCZKA, *Institute of Nuclear Research, Warsaw, Poland*
S. ULAM, *University of Colorado, U.S.A.*

Assistant Editor:

J. C. CORTET, *Université de Dijon, France*

Editorial Board:

H. ARAKI, *Kyoto University, Japan*
A. O. BARUT, *University of Colorado, U.S.A.*
J. P. ECKMANN, *Institut de Physique Théorique, Geneva, Switzerland*
L. FADDEEV, *Steklov Institute of Mathematics, Leningrad, U.S.S.R.*
C. FRONSDAL, *UCLA, Los Angeles, U.S.A.*
I. M. GELFAND, *Moscow State University, U.S.S.R.*
L. GROSS, *Cornell University, U.S.A.*
A. JAFFE, *Harvard University, U.S.A.*
J. P. JURZAK, *Université de Dijon, France*
M. KAC, *The Rockefeller University, New York, U.S.A.*
A. A. KIRILLOV, *Moscow State University, U.S.S.R.*
B. KOSTANT, *M.I.T., U.S.A.*
A. LICHNEROWICZ, *Collège de France, France*
E. H. LIEB, *Princeton University, U.S.A.*
B. NAGEL, *K.T.H., Stockholm, Sweden*
J. NIEDERLE, *Institute of Physics CSAV, Prague, Czechoslovakia*
C. PIRON, *Institut de Physique Théorique, Geneva, Switzerland*
A. SALAM, *International Center for Theoretical Physics, Trieste, Italy*
I. E. SEGAL, *M.I.T., U.S.A.*
D. STERNHEIMER, *Collège de France, France*
E. C. G. SUDARSHAN, *University of Texas, U.S.A.*

VOLUME 2

Non-Relativistic Quantum Dynamics

by

W. O. Amrein

Department of Theoretical Physics,
University of Geneva, Switzerland

D. Reidel Publishing Company

Dordrecht, Holland / Boston, U.S.A. / London, England

Library of Congress Cataloging in Publication Data

CIP

Amrein, Werner O.
Non-relativistic quantum dynamics.

(Mathematical physics studies ; v. 2)
Bibliography: p.
Includes indexes.
1. Quantum theory. 2. Operator theory. I. Title.
II. Series.
QC174.12.A48 530.1′2 81-10704
ISBN 90-277-1324-3 (pbk.) AACR2

Published by D. Reidel Publishing Company
P.O. Box 17, 3300 AA Dordrecht, Holland

Sold and distributed in the U.S.A. and Canada
by Kluwer Boston Inc.,
190 Old Derby Street, Hingham, MA 02043, U.S.A.

In all other countries, sold and distributed
by Kluwer Academic Publishers Group,
P.O. Box 322, 3300 AH Dordrecht, Holland

D. Reidel Publishing Company is a member of the Kluwer Group

Printed in The Netherlands

TABLE OF CONTENTS

PREFACE

The bulk of known results in spectral and scattering theory for Schrödinger operators has been derived by time-independent (also called stationary) methods, which make extensive use of resolvent estimates and the spectral theorem. In very recent years there has been a partial shift of emphasis from the time-independent to the time-dependent theory, due to the discovery of new, essentially time-dependent proofs of a fair number of the principal results such as asymptotic completeness, absence of singularly continuous spectrum and properties of scattering cross sections. These new time-dependent arguments are somewhat simpler than the stationary ones and at the same time considerably closer to physical intuition, in that they are based on a rather detailed description of the time evolution of states in configuration space (whence the designation "geometric methods" used by some authors).

It seemed interesting to me to present some of these new methods from a strictly time-dependent point of view, by considering as the basic mathematical object strongly continuous unitary one-parameter groups and avoiding the spectral theorem completely. The present volume may be viewed as an essay in this spirit. It is an extended version of a course taught in 1979 at the University of Geneva to undergraduate students enrolled in mathematical physics. The text is essentially self-contained, inasmuch as we introduce and develop the basic mathematical tools in the first three chapters, namely Hilbert space and linear operators in Chapter 1, self-adjoint operators in Chapter 2 and Hilbert-Schmidt and compact operators in Chapter 3. Chapter 4 is devoted to a discussion of strongly continuous unitary one-parameter groups and Chapter 5 to asymptotic properties of such groups for large values of the parameter, which leads to various results on spectral theory and wave operators. Chapter 6 contains a short introduction to scattering theory and some discussion of scattering cross sections. An Appendix and the Notes contain some supplementary material.

The presentation is such that we always discuss first the mathematical structure in an abstract form (but restricted to Hilbert space theory) and then apply it to Schrödinger operators in $L^2(\mathbb{R}^\nu)$. The aim is not so much to prove the most general known results but rather to insist on the mathematical methods and the physical meaning. We have however included a discussion of Schrödinger operators with strongly singular potentials. These rather technical developments are confined to Section 2.5 and may be omitted in the first reading, although some familiarity with the principal results will be useful in parts of Chapters 5 and 6.

Finally we should point out that the time-dependent approach does not completely supplement the stationary methods, in that a certain number of the finer results have been obtained only by using the latter.

It is a pleasure to thank my students for their continual interest in this topic, Arne Jensen and David Pearson for profitable discussions, Anne-Marie Berthier for a very useful idea, the editors for their promptness and, above all, Francine Nicole for skillfully typing the difficult manuscript.

CHAPTER 1 : LINEAR OPERATORS IN HILBERT SPACE

In Section 1.1 we give the definition and some elementary properties of a separable Hilbert space, with L^2-spaces as examples. Section 1.2 contains the basic notions about linear operators and a characterization of isometric and unitary operators. Section 1.3 is devoted to a discussion of the Riemann integral of vector-valued and operator-valued functions.

Throughout these notes, a statement or an equation bearing double signs is meant to hold separately for the upper sign and for the lower sign. $\bar{\alpha}$ denotes the complex-conjugate of the number $\alpha \in \mathbb{C}$ and δ_{ij} the Kronecker symbol, i.e. $\delta_{ij} = 1$ if $i = j$ and $\delta_{ij} = 0$ if $i \neq j$.

1.1 Hilbert Space

A complex Hilbert space is a set H of elements f,g,h,... called vectors satisfying the following three axioms :

(H1) H is a linear vector space over the field $\mathbb{C}$ of complex numbers : Whenever $f,g \in H$ and $\alpha,\beta \in \mathbb{C}$, then $f + \alpha g$ is an element of H, and

$$f + g = g + f, \quad (f + g) + h = f + (g + h), \tag{1.1}$$

$$\alpha(f + g) = \alpha f + \alpha g, \quad (\alpha + \beta)f = \alpha f + \beta f, \tag{1.2}$$

$$\alpha(\beta f) = (\alpha\beta)f, \quad 1 \cdot f = f, \tag{1.3}$$

and there exists a vector θ, called the zero vector, such that

$$f + \theta = f \quad \text{and} \quad 0 \cdot f = \theta \quad \text{for all} \quad f \in H. \tag{1.4}$$

(H2) There exists on H a positive definite <u>scalar product</u>, i.e. a mapping from $H \times H$ to $\mathbb{C}$, denoted by $(\cdot,\cdot)$, such that for all $f,g,h \in H$ and $\alpha,\beta \in \mathbb{C}$:

$$(g,f) = \overline{(f,g)}, \tag{1.5}$$

$$(f,\alpha g + \beta h) = \alpha(f,g) + \beta(f,h), \tag{1.6}$$

$$(f,f) \geq 0, \text{ and } (f,f) = 0 \text{ only for } f = \theta. \tag{1.7}$$

The scalar product induces a metric on H. The distance $d(f,g)$ between two vectors f and g is $d(f,g) = \|f-g\|$, where the <u>norm</u> $\|\cdot\|$ is defined as

$$\|f\| := (f,f)^{1/2}. \tag{1.8}$$

(H3) H is <u>complete</u> with respect to the norm (1.8) : If $\{f_n\}$ is a Cauchy sequence in H, i.e. such that $\|f_n - f_m\| \to 0$ as $m,n \to \infty$, there exists a vector f in H such that $\|f - f_n\| \to 0$ as $n \to \infty$.

<u>Definition</u> : A Hilbert space H is <u>separable</u> if it has a countable orthonormal basis, i.e. if there is a sequence $\{e_i\}$ in H, $i = 1,2,\ldots$, such that

$$(e_i,e_j) = \delta_{ij}$$

and such that each $f \in H$ is the limit, in the norm (1.8), of a sequence of finite linear combinations of the vectors $\{e_i\}$. (In other words, there are $\alpha_i^n \in \mathbb{C}$ such that $\|f - f_n\| \to 0$ as $n \to \infty$, with $f_n = \sum_{i=1}^n \alpha_i^n e_i$.) The cardinal number of the set $\{e_i\}$ is called the <u>dimension</u> of H.

Throughout these lectures H is assumed to be a separable complex Hilbert space.

The scalar product leads naturally to a notion of orthogonality in H. The vectors f and g are said to be orthogonal to each other, written $f \perp g$, if $(f,g) = 0$. If M is a subset of H, we write $f \perp M$ if $f \perp g$ for each $g \in M$. The orthogonal complement $M^\perp$ of M is defined as

$$M^\perp = \{f \in H \mid f \perp M\}. \tag{1.9}$$

Two simple consequences of the axioms of Hilbert space are the following inequalities :

i) the Schwarz inequality

$$|(f,g)| \leq \|f\| \, \|g\|, \tag{1.10}$$

ii) the triangle inequality

$$\|f \pm g\| \leq \|f\| + \|g\|. \tag{1.11}$$

Proof : To prove (1.10), one notices that $(f + \alpha g, f + \alpha g)$ is a positive definite quadratic form in α, and then sets $\alpha =$ $= (g,f)/(g,g)$. The details are left to the reader. (1.11) follows from (1.10) :

$$\|f \pm g\|^2 = (f \pm g, f \pm g) = (f,f) \pm (f,g) \pm (g,f) + (g,g)$$

$$\leq \|f\|^2 + 2|(f,g)| + \|g\|^2 \leq \|f\|^2 + 2\|f\| \, \|g\| + \|g\|^2$$

$$= (\|f\| + \|g\|)^2. \qquad \square$$

Since $(\alpha + \beta)^2 \leq 2\alpha^2 + 2\beta^2$, the preceding inequality also implies that

$$\|f \pm g\|^2 \leq 2\|f\|^2 + 2\|g\|^2. \tag{1.12}$$

In a similar way, using $(\sum_{k=1}^n \alpha_k)^2 \leq n \sum_{k=1}^n \alpha_k^2$ if $\alpha_k \in \mathbb{R}$, one gets

$$\|\sum_{k=1}^{n} f_k\|^2 \leq n \sum_{k=1}^{n} \|f_k\|^2. \qquad (1.13)$$

We shall also use the polarization identity which allows one to express the scalar product in terms of the norm, viz.

$$4(f,g) = \|f+g\|^2 - \|f-g\|^2 - i\|f+ig\|^2 + i\|f-ig\|^2. \qquad (1.14)$$

This identity is easily verified by using the definition (1.8) of the norm and the linearity of the scalar product as in the proof of (1.11).

Another simple consequence of (1.11) is the inequality

$$|\|f\| - \|g\|| \leq \|f-g\|. \qquad (1.15)$$

To prove it, we may assume that $\|g\| \leq \|f\|$ and then get that

$$\|f\| - \|g\| = \|f-g+g\| - \|g\| \leq (\|f-g\| + \|g\|) - \|g\| = \|f-g\|.$$

There are two types of convergence in Hilbert space, namely strong convergence and weak convergence :

i) A sequence $\{f_n\}$ in H is said to converge strongly to a vector f in H, written s-lim $f_n = f$, if $\lim\|f-f_n\| = 0$ as $n \to \infty$.

ii) A sequence $\{f_n\}$ in H is said to converge weakly to a vector f in H, written w-lim $f_n = f$, if $\lim(g,f_n) = (g,f)$ as $n \to \infty$, for each $g \in H$.

The following theorem relates these two types of convergence. Together with Example 1.2 it shows that weak convergence is a weaker notion of convergence than strong convergence.

Proposition 1.1 : Let $\{f_n\}$, $f \in H$. Then

$$\text{s-}\lim_{n\to\infty} f_n = f \iff \text{w-}\lim_{n\to\infty} f_n = f \text{ and } \lim_{n\to\infty}\|f_n\| = \|f\|.$$

Proof : (i) Suppose $s\text{-}\lim f_n = f$. By the Schwarz inequality, one has for each $g \in H$

$$|(g,f_n) - (g,f)| = |(g,f_n - f)| \leq \|g\| \, \|f_n - f\| \to 0 \text{ as } n \to \infty,$$

i.e. $w\text{-}\lim f_n = f$.

The inequality (1.15) implies that

$$| \|f_n\| - \|f\| | \leq \|f_n - f\|,$$

which converges to zero as $n \to \infty$, because of the hypothesis that $s\text{-}\lim f_n = f$. Hence $\lim \|f_n\| = \|f\|$ as $n \to \infty$.

(ii) To prove the converse, we notice that

$$\|f - f_n\|^2 = \|f\|^2 + \|f_n\|^2 - (f,f_n) - (f_n,f).$$

If $w\text{-}\lim f_n = f$ and $\lim \|f_n\| = \|f\|$, the r.h.s. converges to $\|f\|^2 + \|f\|^2 - \|f\|^2 - \|f\|^2 = 0$, i.e. $s\text{-}\lim f_n = f$. □

Example 1.2 : Let $\{f_n\}_{n=1}^{\infty}$ be an infinite orthonormal sequence (in an infinite-dimensional Hilbert space H), i.e. $(f_i,f_j) = \delta_{ij}$. Then $w\text{-}\lim f_n = \theta$, since for any $g \in H$,

$$\|g\|^2 \geq \sum_{n=1}^{\infty} |(f_n,g)|^2, \qquad (1.16)$$

hence $\lim(g,f_n) = 0$ as $n \to \infty$. In fact, $\{f_n\}$ may be considered to be part of an orthonormal basis of H, and for an orthonormal basis $\{f_n\}$, (1.16) holds with the equality sign and simply expresses the fact that (f_n,g) is the coefficient of f_n in the expansion of g with respect to the basis $\{f_n\}$.

Thus each infinite orthonormal sequence $\{f_n\}$ converges weakly to zero. On the other hand $\|f_n\| = 1 \not\to 0$, so that by Lemma 1.1 $\{f_n\}$ cannot be strongly convergent. This may also be seen directly, since $||f_n - f_m||^2 = 2$ if $n \neq m$.

Definition : A subspace of H is a subset M of H that satisfies the axioms (H1), (H2) and (H3). A linear manifold is a subset M of H that satisfies the axioms (H1) and (H2), but not necessarily (H3).

Thus a linear manifold M is a linear vector space; however, the limit of a strong Cauchy sequence $\{f_n\}$ in M may not be in M. One may complete M by adding to it all limits of strong Cauchy sequences in M. The subset of H thus obtained is called the closure of M and denoted by $\bar{M}$. Clearly $\bar{M}$ is a subspace, and it is easy to see that $\bar{M} = (M^{\perp})^{\perp}$. (This and the proof of Lemmas 1.4 and 1.5 requires the Projection Lemma : If M is a subspace, each $f \in H$ has a unique decomposition into $f = g + h$ with $g \in M$, $h \in M^{\perp}$.)

Problem 1.3 : Let N be an arbitrary subset of H. Then $N^{\perp}$ is a subspace.

Lemma 1.4 : Let M_1 and M_2 be subspaces of a Hilbert space H such that M_1 is infinite-dimensional and M_2 finite-dimensional. Then there is a vector $h \neq \theta$ such that $h \in M_1$ and $h \perp M_2$.

Proof : Let $M_{21} \subset M_1$ be the orthogonal projection of M_2 into M_1 : Each $f \in M_2$ may be written as $f = f_1 + g_1$ with $f_1 \in M_1$, $g_1 \in M_1^{\perp}$. M_{21} is the finite-dimensional subspace of M_1 formed by all vectors f_1. Since $\dim M_1 = \infty$ and $\dim M_{21} \leq \dim M_2 < \infty$, the orthogonal complement of M_{21} in M_1 is infinite-dimensional, and any vector h in this orthogonal complement has the required properties. □

Definition : A subset D of H is dense in H if, for each $f \in H$ and each $\varepsilon > 0$, there is a vector f_ε in D such that $||f - f_\varepsilon|| < \varepsilon$. A subset D_o of H is total in H if its linear span D, i.e. the set of finite linear combinations of vectors in D_o, is dense in H.

An example of a total set is an orthonormal basis $\{e_n\}$ of H. The linear span of an orthonormal basis in an infinite-dimensional Hilbert space is a dense linear manifold which is not a subspace.

Lemma 1.5 : $\mathcal{D}_o$ is total in H if and only if $f \perp \mathcal{D}_o$ implies $f = \theta$.

Proof : (i) Assume $\mathcal{D}_o$ to be total, and let $\mathcal{D}$ be its linear span. Suppose $f \perp \mathcal{D}_o$ and $f \neq \theta$. Then $f \perp \mathcal{D}$, since each $g \in \mathcal{D}$ has the form $g = \sum_{i=1}^n \alpha_i g_i$ with $g_i \in \mathcal{D}_o$ and $n < \infty$, hence $(f,g) = \sum_{i=1}^n \alpha_i (f,g_i) = 0$.

Let $\varepsilon > 0$. Choose $g_\varepsilon \in \mathcal{D}$ such that $||f - g_\varepsilon|| < \varepsilon ||f||^{-1}$. Since $(f,g_\varepsilon) = 0$, we have by (1.10)

$$||f||^2 = (f,f) = |(f,f - g_\varepsilon)| \leq ||f|| \, ||f - g_\varepsilon|| < \varepsilon .$$

Since ε is arbitrary, $||f||^2 = 0$, hence $f = \theta$ by (1.7).

(ii) Assume $f \perp \mathcal{D}_o$ implies $f = \theta$. It follows that $\mathcal{D}_o^\perp = \mathcal{D}^\perp = \{\theta\}$, hence by the Projection Lemma $\bar{\mathcal{D}} = \{\theta\}^\perp = H$. □

Remark 1.6 : A strong Cauchy sequence in H converges by (H3). One can also show that each weak Cauchy sequence converges weakly to a limit vector f in H. Lemma 1.5 implies for both cases that the limit is unique. In fact, since strong convergence implies weak convergence, it suffices to prove uniqueness in the latter case. So suppose w-lim $f_n = f$ and w-lim $f_n = h$. Then one has for each $g \in H$:

$$(g,f - h) = (g,f) - (g,h) = \lim_{n\to\infty}[(g,f_n) - (g,f_n)] = 0.$$

Thus $f - h$ is orthogonal to H, hence $f - h = \theta$ or $f = h$ by Lemma 1.5.

In these lectures we shall deal with three types of Hilbert spaces, namely ℓ^2, L^2 and B_2. The space B_2 will be introduced in Chapter 3. The space ℓ^2 is the set of all infinite sequences

$\alpha = \{\alpha_1, \alpha_2, \alpha_3, \ldots\}$ of complex numbers satisfying $\sum_{i=1}^{\infty} |\alpha_i|^2 < \infty$, with the scalar product ($\beta = \{\beta_i\}$)

$$(\alpha, \beta) = \sum_{i=1}^{\infty} \overline{\alpha_i} \beta_i .$$

Every infinite-dimensional separable Hilbert space H is isomorphic to ℓ^2.

Let (M, μ) be a measure space (see the Appendix for more details). One considers μ-measurable functions $f: M \to \mathbb{C}$ such that

$$\int_M |f(s)|^2 d\mu(s) < \infty.$$

These functions form a linear vector space if one defines

$$(f+g)(s) = f(s) + g(s), \qquad (\alpha f)(s) = \alpha f(s),$$

and one may introduce a scalar product by

$$(f,g) = \int_M \overline{f(s)} g(s) d\mu(s).$$

Two functions f_1 and f_2 are said to be equivalent if they differ at most on a set of μ-measure zero, i.e. if

$$\int_M |f_1(s) - f_2(s)|^2 d\mu(s) = 0.$$

The set of equivalence classes of such square-integrable functions is a Hilbert space called $L^2(M; d\mu)$.

We shall mostly deal with the case where M is an open subset Δ of Euclidean ν-space $\mathbb{R}^\nu$ and $d\mu$ is Lebesgue measure $d^\nu x$ on Δ. The corresponding L^2-space will be denoted by $L^2(\Delta)$.

If Δ is an open subset of $\mathbb{R}^\nu$, we denote by $C_o^\infty(\Delta)$ the set of all infinitely differentiable functions from $\mathbb{R}^\nu$ to $\mathbb{C}$ having compact support in Δ. Thus $f:\mathbb{R}^\nu \to \mathbb{C}$ belongs to $C_0^\infty(\Delta)$ if and only if it is infinitely differentiable and supp f, the closure of the set $\{\underline{x}\in\mathbb{R}^\nu | f(\underline{x}) \neq 0\}$, is bounded and contained in Δ. We shall often use the fact that $C_0^\infty(\Delta)$ is dense in $L^2(\Delta)$. More details on this may be found in the Appendix, which also contains the definition and properties of the spaces $L^p(\Delta)$ for $p \neq 2$.

Let us consider in particular $L^2(\mathbb{R}^\nu)$, i.e. $\Delta = \mathbb{R}^\nu$. We shall frequently refer to the <u>Fourier transformation</u> F on this space. It is first defined on $L^2(\mathbb{R}^\nu) \cap L^1(\mathbb{R}^\nu)$ by

$$(Ff)(\underline{k}) \equiv \tilde{f}(\underline{k}) := (2\pi)^{-\nu/2} \int e^{-i\underline{k}\cdot\underline{x}} f(\underline{x}) d^\nu x, \tag{1.17}$$

where $\underline{k}\in\mathbb{R}^\nu$ and $\underline{k}\cdot\underline{x} = \sum_{i=1}^{\nu} k_i x_i$. If $f\in L^2(\mathbb{R}^\nu)$ but $f\notin L^1(\mathbb{R}^\nu)$, the integral in (1.17) does not have a pointwise meaning. One then defines Ff by approximating f by a sequence $\{f_N\}$ belonging to the dense set $L^2(\mathbb{R}^\nu) \cap L^1(\mathbb{R}^\nu)$. For instance one may take

$$f_N(\underline{x}) = \begin{cases} f(\underline{x}) & \text{if } |\underline{x}| \le N \\ 0 & \text{if } |\underline{x}| > N, \end{cases}$$

where $|\underline{x}| = (\underline{x}\cdot\underline{x})^{1/2}$ is the Euclidean length of $\underline{x}$. By Lemma A.1, $f_N\in L^1(\mathbb{R}^\nu) \cap L^2(\mathbb{R}^\nu)$. Now <u>$F$ is unitary</u>, i.e.

$$\int |\tilde{f}(\underline{k})|^2 d^\nu k = \int |f(\underline{x})|^2 d^\nu x, \tag{1.18}$$

so that, if $M > N$,

$$||Ff_N - Ff_M||^2 = ||f_N - f_M||^2 = \int_{N\le|\underline{x}|\le M} |f(\underline{x})|^2 d^\nu x,$$

which converges to zero as $M,N \to \infty$, since $f\in L^2(\mathbb{R}^\nu)$. Hence one may define Ff as the strong limit in $L^2(\mathbb{R}^\nu)$ of Ff_N :

$$Ff = \operatorname*{s-lim}_{N\to\infty} Ff_N. \tag{1.19}$$

This is also called limit in the mean and written as

$$(Ff)(\underline{k}) = \text{l.i.m.}(2\pi)^{-\nu/2}\int e^{-i\underline{k}\cdot\underline{x}} f(\underline{x})d^\nu x. \quad (1.20)$$

The unitarity relation (1.18) holds for each $f \in L^2(\mathbb{R}^\nu)$. The Fourier transformation is invertible, and its inverse F^{-1} is given, for $\tilde{f} \in L^2(\mathbb{R}^\nu) \cap L^1(\mathbb{R}^\nu)$, by

$$f(\underline{x}) = (2\pi)^{-\nu/2}\int e^{i\underline{k}\cdot\underline{x}}\tilde{f}(\underline{k})d^\nu k. \quad (1.21)$$

Although the set $\{\tilde{f} | f \in L^2(\mathbb{R}^\nu)\}$ is again $L^2(\mathbb{R}^\nu)$, it is convenient to distinguish the two representations of $L^2(\mathbb{R}^\nu)$, because the variables $\underline{x}$ and $\underline{k}$ have different physical interpretations in quantum mechanics. Multiplication of $f(\underline{x})$ by x_i corresponds to the i-th component of the position operator, and multiplication of $\tilde{f}(\underline{k})$ by k_i to the i-th component of the momentum operator, see Section 2.3. We shall therefore denote the set of functions $\{f(\underline{x}) | f \in L^2(\mathbb{R}^\nu)\}$ by $L^2(\mathbb{R}^\nu)$ and the set $\{\tilde{f}(\underline{k}) | f \in L^2(\mathbb{R}^\nu)\}$ of their Fourier transforms by $\tilde{L}^2(\mathbb{R}^\nu)$. In other words, we do not consider $L^2(\mathbb{R}^\nu)$ as an abstract space but rather as the set of square-integrable quantum-mechanical wave functions defined on ν-dimensional configuration space.

1.2 Linear Operators

Let H and H' be two Hilbert spaces. A linear operator from H to H' is a couple $\{D(A),A\}$, where $D(A)$ is a linear manifold in H and A a linear mapping from $D(A)$ into H'. In other words, to each $f \in D(A)$ there is associated a vector $Af \in H'$ in such a way that

$$A(f+\alpha g) = Af + \alpha\cdot Ag \quad \text{for} \quad f,g \in D(A), \quad \alpha \in \mathbb{C}.$$

The linear manifold $D(A)$ is called the domain of A. It is customary to denote a linear operator $\{D(A),A\}$ simply by A.

We shall mostly deal with linear operators acting in one Hilbert space. For this reason we now assume that $\mathcal{H}' = \mathcal{H}$ and study linear operators A from $D(A) \subseteq \mathcal{H}$ into $\mathcal{H}$. We add a few more definitions : The <u>range</u> $\mathcal{R}(A)$ of A is the image of D(A) under A, i.e.

$$\mathcal{R}(A) := AD(A) = \{f \in \mathcal{H} \mid f = Ag \quad \text{for some} \quad g \in D(A)\}.$$

The <u>null space</u> N(A) of A is the linear submanifold of D(A) mapped onto the zero vector :

$$N(A) = \{f \in D(A) \mid Af = \theta\}.$$

Two operators A and B are <u>equal</u>, written A = B, if D(A) = = D(B) and Af = Bf for each $f \in D(A)$. B is said to be an <u>extension</u> of A if $D(A) \subseteq D(B)$ and Bf = Af for each $f \in D(A)$. In this case A is also called the <u>restriction</u> of B to D(A), and one writes $B \supseteq A$ or $A \subseteq B$.

The <u>sum</u> A + B of two operators A and B is defined as follows : $D(A+B) = D(A) \cap D(B)$, and $(A+B)f = Af + Bf$ for $f \in D(A+B)$. (Notice that it may happen that $D(A+B) = \{\theta\}$.)

A is <u>bounded</u> if there is a number $M < \infty$ such that

$$||Af|| \leq M||f|| \qquad \forall\, f \in D(A). \tag{1.22}$$

If A is bounded, one defines the <u>norm</u> $||A||$ of A by

$$||A|| = \sup_{\theta \neq f \in D(A)} \frac{||Af||}{||f||}. \tag{1.23}$$

One then has

$$||Af|| \leq ||A||\,||f|| \qquad \forall\, f \in D(A). \tag{1.24}$$

We denote by $\mathcal{B}(\mathcal{H})$ the set of all bounded operators A on $\mathcal{H}$ for which

$D(A) = \mathcal{H}$. For $A, B \in \mathcal{B}(\mathcal{H})$, one immediately finds from (1.24) and (1.11) that $AB \in \mathcal{B}(\mathcal{H})$, $(A+B) \in \mathcal{B}(\mathcal{H})$ and

$$||AB|| \leq ||A|| \, ||B||, \tag{1.25}$$

$$||A+B|| \leq ||A|| + ||B||. \tag{1.26}$$

Proposition 1.7 : Let A be bounded. Then A has a unique bounded extension $\bar{A}$ with domain $D(\bar{A}) = \overline{D(A)}$ (the closure of $D(A)$) such that $||\bar{A}|| = ||A||$. $\bar{A}$ is called the closure of A. In particular, if A is bounded and densely defined, its closure $\bar{A}$ belongs to $\mathcal{B}(\mathcal{H})$.

Proof : Let $g \in \overline{D(A)}$. There is a sequence $\{g_n\} \in D(A)$ such that $g =$ $=$ s-lim g_n. Since A is bounded, $\{Ag_n\}$ is also a strong Cauchy sequence :

$$||Ag_n - Ag_m|| \leq ||A|| \, ||g_n - g_m|| \to 0 \quad \text{as} \quad m,n \to \infty.$$

Let $f =$ s-lim Ag_n, and set $\bar{A}g = f$. It is easy to check that this definition is independent of the particular sequence $\{g_n\}$ converging to g, hence $\bar{A}$ is well defined.

Clearly $\bar{A}$ is an extension of A : if $g \in D(A)$, it suffices to take $g_n = g$. Furthermore, by applying twice Proposition 1.1,

$$||\bar{A}g|| = \lim_{n\to\infty} ||Ag_n|| \leq ||A|| \lim_{n\to\infty} ||g_n|| = ||A|| \, ||g||.$$

Hence

$$||\bar{A}|| = \sup_{\theta \neq g \in D(\bar{A})} \frac{||\bar{A}g||}{||g||} \leq ||A||. \tag{1.27}$$

In fact one has equality in (1.27), since the supremum over all $g \neq \theta$ in the subset $D(A)$ of $D(\bar{A})$ is already equal to $||A||$. We leave it to the reader to check that $\bar{A}$ is the only operator having all required properties. □

Lemma 1.8 : Let $\{A_n\} \in \mathcal{B}(\mathcal{H})$ be such that $\|A_n\| \le M < \infty$ for all $n = 1,2,\ldots$. Let $\mathcal{M}$ be a subspace of $\mathcal{H}$ and suppose there is a subset $\mathcal{D}_0$ of $\mathcal{M}$ which is total in $\mathcal{M}$ and such that, for each $f \in \mathcal{D}_0$, the sequence $\{A_n f\}$ is strongly Cauchy. Then s-lim $A_n g$ as $n \to \infty$ exists for each $g \in \mathcal{M}$. If furthermore $A \in \mathcal{B}(\mathcal{H})$ is such that s-lim $A_n f = Af$ for all $f \in \mathcal{D}_0$, then s-lim $A_n g = Ag$ for each $g \in \mathcal{M}$.

Proof : Clearly s-lim $A_n f$ exists for each f in the linear span $\mathcal{D}$ of $\mathcal{D}_0$. Now, if $g \in \mathcal{M}$ and $\varepsilon > 0$, we may choose a vector $f \in \mathcal{D}$ such that $\|g - f\| < \varepsilon/(3M)$. Then, by the triangle inequality,

$$\|A_n g - A_m g\| \le \|A_n(g - f)\| + \|A_m(f - g)\| + \|A_n f - A_m f\|$$

$$\le 2M\|g - f\| + \|A_n f - A_m f\| \le 2\varepsilon/3 + \|A_n f - A_m f\| .$$

Since $\{A_n f\}$ is strongly Cauchy, $\|A_n f - A_m f\| < \varepsilon/3$ provided that $m,n \ge n_0(\varepsilon)$. Hence $\|A_n g - A_m g\| < \varepsilon$ whenever $m,n \ge n_0(\varepsilon)$, i.e. $\{A_n g\}$ is strongly Cauchy.

The proof of the second assertion is similar; it suffices to replace A_m by A in the above inequality. □

In the preceding lemma we have introduced the notion of strong convergence of a sequence $\{A_n\}$ of operators in $\mathcal{B}(\mathcal{H})$, viz.

$$\text{s-}\lim_{n\to\infty} A_n = A \iff \lim_{n\to\infty}\|Af - A_n f\| = 0 \quad \forall f \in \mathcal{H} .$$

We shall need two other types of convergence in $\mathcal{B}(\mathcal{H})$, namely weak convergence and uniform convergence. These are defined as follows :

$$\text{w-}\lim_{n\to\infty} A_n = A \iff \lim_{n\to\infty}(f, A_n g) = (f, Ag) \quad \forall f,g \in \mathcal{H},$$

and

$$\text{u-}\lim_{n\to\infty} A_n = A \iff \lim_{n\to\infty}\|A - A_n\| = 0.$$

Uniform convergence is also called convergence in the operator norm.

By Proposition 1.1, strong convergence of a sequence of operators $\{A_n\}$ to A implies weak convergence of $\{A_n\}$ to A. Similarly uniform convergence implies strong convergence : If $||A - A_n|| \to 0$ and $f \in \mathcal{H}$, then $||Af - A_n f|| \leq ||A - A_n|| \, ||f|| \to 0$.

Lemma 1.9 : Let $A, B, A_n, B_n \in \mathcal{B}(\mathcal{H})$ and assume that $\text{s-lim } A_n = A$ and $\text{s-lim } B_n = B$ as $n \to \infty$. Then $\text{s-lim } A_n B_n = AB$ as $n \to \infty$.

Proof : We use the fact that a strongly convergent sequence in $\mathcal{B}(\mathcal{H})$ is bounded (the uniform boundedness principle). Thus there is $M < \infty$ such that $||A_n|| \leq M$ for all n. In all subsequent applications of Lemma 1.9 the boundedness of $\{A_n\}$ is evident.

Let $f \in \mathcal{H}$. Then

$$||ABf - A_n B_n f|| \leq ||(A - A_n)Bf|| + ||A_n(B - B_n)f||$$

$$\leq ||(A - A_n)Bf|| + M||(B - B_n)f||,$$

which converges to zero as $n \to \infty$. □

Remark 1.10 : Lemma 1.9 remains true if strong convergence is replaced everywhere by uniform convergence, but not so if strong convergence is replaced by weak convergence.

A linear operator A is said to be invertible if $N(A) = \{\theta\}$. In that case, the inverse A^{-1} of A is the following operator : $D(A^{-1}) = \mathcal{R}(A)$, and $A^{-1}(Af) = f$ for $f \in D(A)$. We shall use the following theorem about invertibility. In its statement we use the identity operator I, defined by $D(I) = \mathcal{H}$ and $If = f$ for each $f \in \mathcal{H}$.

Proposition 1.11 : Let $A \in \mathcal{B}(\mathcal{H})$ be such that $||A|| < 1$. Then (a)

I - A is invertible, (b) $\mathcal{R}(I - A) = \mathcal{H}$, (c) $(I - A)^{-1} \in \mathcal{B}(\mathcal{H})$ and

$$\|(I - A)^{-1}\| \leq (1 - \|A\|)^{-1}, \tag{1.28}$$

(d) $(I - A)^{-1}$ is given by the following uniformly convergent series (called the Neumann series) :

$$(I - A)^{-1} = I + A + A^2 + \ldots = \sum_{n=0}^{\infty} A^n. \tag{1.29}$$

Proof : The series in (1.29) is uniformly convergent, since (1.25) and (1.26) imply that for $n > m$

$$\|\sum_{k=m}^{n} A^k\| \leq \|A^m\| \sum_{k=0}^{n-m} \|A^k\|$$

$$\leq \|A\|^m \sum_{k=0}^{\infty} \|A\|^k = \|A\|^m (1 - \|A\|)^{-1}, \tag{1.30}$$

which converges to zero as $m \to \infty$. Thus the series defines a linear operator B with $D(B) = \mathcal{H}$. Furthermore $B \in \mathcal{B}(\mathcal{H})$, since

$$\|B\| = \|\sum_{k=0}^{\infty} A^k\| \leq (1 - \|A\|)^{-1} < \infty.$$

Now notice that

$$(I + A + A^2 + \ldots + A^n)(I - A) = I - A^{n+1}. \tag{1.31}$$

But $\|A^{n+1}\| \leq \|A\|^{n+1} \to 0$ as $n \to \infty$, since $\|A\| < 1$. Hence (1.31) implies that

$$(\sum_{k=0}^{\infty} A^k)(I - A) = I.$$

Hence $(I - A)$ is invertible, and $(I - A)^{-1} = B$ is given by (1.29). A similar calculation shows that $(I - A)(\sum_{k=0}^{\infty} A_k) = I$, hence $\mathcal{R}(I - A) = \mathcal{H}$. □

We next introduce the <u>adjoint</u> operator A^* of A. To define A^*, we must assume that D(A) is dense in $\mathcal{H}$. We first give the domain $D(A^*)$ of the adjoint :

$g \in D(A^*)$ if there is a vector $g^* \in \mathcal{H}$ such that

$$(Af,g) = (f,g^*) \qquad \forall\, f \in D(A). \tag{1.32}$$

Note that there is at most one such vector g^* : If in addition to (1.32), $(Af,g) = (f,g_1^*)$ for all $f \in D(A)$, then $g^* - g_1^*$ is orthogonal to the dense set D(A), hence $g^* = g_1^*$ by Lemma 1.5.

The linear mapping $A^*: D(A^*) \to \mathcal{H}$ is now defined by setting $A^*g = g^*$, for $g \in D(A^*)$. Thus one has

$$(Af,g) = (f,A^*g) \text{ for all } f \in D(A) \text{ and } g \in D(A^*). \tag{1.33}$$

We state without proof that, if $A \in \mathcal{B}(\mathcal{H})$, then A^* is also in $\mathcal{B}(\mathcal{H})$, and

$$\|A^*\| = \|A\|. \tag{1.34}$$

Some further properties of the adjoint are given below. Their proof is simple and follows immediately from the above definition of A^*.

<u>Problem 1.12</u> : (a) If $A,B \in \mathcal{B}(\mathcal{H})$, then

$$(AB)^* = B^*A^*, \tag{1.35}$$

$$A^{**} := (A^*)^* = A. \tag{1.36}$$

(b) If D(A) and $D(A^*)$ are dense, then

$$A \subseteq A^{**}. \tag{1.37}$$

(c) If D(A), D(B) are dense and $\mathcal{R}(B) \subseteq D(A)$, then

$$(AB)^* \supseteq B^*A^*. \tag{1.38}$$

(The product of two operators S and T is defined by $(ST)f = S(Tf)$ with $D(ST) = \{f \in D(T) \mid Tf \in D(S)\}$.)

(d) If $D(A)$ is dense, $B \in \mathcal{B}(\mathcal{H})$ and $\alpha \neq 0$, then $D((\alpha A + B)^*) = D(A^*)$ and $(\alpha A + B)^* = \bar{\alpha}A^* + B^*$.

(e) If $D(A)$ is dense and $A \subseteq B$, then $B^* \subseteq A^*$.

Proposition 1.13 : If $D(A)$ is dense, then

$$N(A^*) = \mathcal{R}(A)^{\perp}. \tag{1.39}$$

Proof : (i) Let $g \in \mathcal{R}(A)^{\perp}$. Then $(Af,g) = 0 \ \forall f \in D(A)$. Hence $g \in D(A^*)$, and $A^*g = g^* = \theta$, i.e. $g \in N(A^*)$.

(ii) Let $g \in N(A^*)$. Then by (1.33)

$$(Af,g) = (f,A^*g) = (f,\theta) = 0 \qquad \forall f \in D(A),$$

hence $g \perp \mathcal{R}(A)$. □

Definition : The linear operator A is closed if, whenever $\{f_n\} \in D(A)$, $\{f_n\}$ is strongly Cauchy and $\{Af_n\}$ is strongly Cauchy, one has $\text{s-lim}\, f_n \in D(A)$ and $A(\text{s-lim}\, f_n) = \text{s-lim}\, Af_n$.

Proposition 1.14 : Let $D(A)$ be dense. Then A^* is closed.

Proof : Assume $\{g_n\} \in D(A^*)$, $\text{s-lim}\, g_n = g$ and $\text{s-lim}\, A^*g_n = h$. We must show that $g \in D(A^*)$ and $A^*g = h$. For this, let $f \in D(A)$. Then, by (1.33),

$$(Af,g) = \lim_{n\to\infty}(Af,g_n) = \lim_{n\to\infty}(f,A^*g_n) = (f,h).$$

Thus, by the definition of the adjoint, our assertion is proven. □

Problem 1.15 : If A is closed, then N(A) is a subspace.

An orthogonal projection operator (or in short a projection) is a linear operator E satisfying $D(E) = \mathcal{H}$ and

$$E^2 = E = E^*. \quad (1.40)$$

We set

$$\mathcal{M}(E) = \{f \in \mathcal{H} | Ef = f\}. \quad (1.41)$$

It is easy to see that $\mathcal{M}(E)$ is a subspace. Furthermore, if $g \perp \mathcal{M}(E)$, we have for any $h \in \mathcal{H}$

$$(Eg,h) = (g,E^*h) = (g,Eh). \quad (1.42)$$

Now $E^2h = Eh$, hence $Eh \in \mathcal{M}(E)$, so that (1.42) implies that $(Eg,h) = 0$. Hence $Eg = \theta$ by Lemma 1.5. This shows that E is nothing but the operation of orthogonal projection of $\mathcal{H}$ onto $\mathcal{M}(E)$.

Problem 1.16 : If E is a non-zero projection, then $||E|| = 1$.

An isometry (or isometric operator) is a linear operator Ω in $\mathcal{B}(\mathcal{H})$ satisfying

$$\Omega^*\Omega = I. \quad (1.43)$$

Proposition 1.17 : Let Ω be an isometry. Then

(a) Ω preserves all scalar products :

$$(\Omega f, \Omega g) = (f,g) \quad \text{for all} \quad f,g \in \mathcal{H}. \quad (1.44)$$

In particular

$$||\Omega f|| = ||f|| \quad \text{for each} \quad f \in \mathcal{H}. \quad (1.45)$$

(b) $\|\Omega\| = 1$.

(c) $\Omega\Omega^*$ is a projection, and $M(\Omega\Omega^*) = R(\Omega)$.

(d) Ω is invertible.

(e) $\Omega^*f = \Omega^{-1}f$ if $f \in R(\Omega)$, and $\Omega^*f = \theta$ if $f \perp R(\Omega)$.

<u>Proof</u> : (a) $(\Omega f, \Omega g) = (f, \Omega^*\Omega g) = (f,g)$.

(b) $$\|\Omega\| = \sup_{\theta \neq f \in H} \frac{\|\Omega f\|}{\|f\|} = \sup_{f \neq \theta} \frac{\|f\|}{\|f\|} = 1, \text{ by } (1.45).$$

(c) Set $F = \Omega\Omega^*$. Then, by (1.35) and (1.36),

$$F^2 = \Omega(\Omega^*\Omega)\Omega^* = \Omega I \Omega^* = \Omega\Omega^* = F,$$

$$F^* = (\Omega\Omega^*)^* = \Omega^{**}\Omega^* = \Omega\Omega^* = F.$$

Thus F is a projection.

Clearly $M(F) \subseteq R(\Omega)$. Conversely, if $g = \Omega f \in R(\Omega)$, then $Fg = \Omega\Omega^*\Omega f = \Omega I f = \Omega f = g$, hence $g \in M(F)$. Thus $M(F) = R(\Omega)$.

(d) By (1.45), $\Omega f = \theta$ implies $\|f\| = 0$, i.e. $f = \theta$. Thus $N(\Omega) = \{\theta\}$, i.e. Ω is invertible.

(e) Let $g = \Omega f$. Then $\Omega^{-1}g = f = \Omega^*\Omega f = \Omega^* g$. On the other hand, if $g \perp R(\Omega)$, then $\Omega^* g = \theta$ by Proposition 1.13. □

The preceding proposition shows that an isometric operator Ω maps the Hilbert space H onto a subspace $M(\Omega\Omega^*)$ while preserving the length of vectors and the angles between vectors. A special case is a <u>unitary operator</u> U for which $M(UU^*) = H$. Thus U is unitary if it is isometric and $F \equiv UU^* = I$; in other words U is unitary if

$$U^*U = I \quad \text{and} \quad UU^* = I. \tag{1.46}$$

In this case $U^* = U^{-1}$ on all of H (cf. Proposition 1.17 (e)).

Lemma 1.18 : Let U be unitary, let N be a subset of H such that both U and U* leave N invariant. Then U and U* leave the subspace M spanned by N (the closure of the linear span of N) and its orthogonal complement $M^{\perp}$ invariant, and the restriction U_o of U to M (viewed as an operator in M) is unitary.

Proof : We have $M^{\perp} = N^{\perp}$. If $f \in M^{\perp}$, $g \in N$, then $(Uf,g) = (f,U^*g) = 0$, since $U^*g \in N$. Thus $Uf \in M^{\perp}$, i.e. U leaves $M^{\perp}$ invariant. Similarly one gets that U* leaves $M^{\perp}$ invariant.

Now if $f \in M^{\perp}$, $g \in M$, then $(f,Ug) = (U^*f,g) = 0$, since $U^*f \in M^{\perp}$. Thus U leaves M invariant. Similarly one sees that U* leaves M invariant.

Denote by $(U^*)_o$ the restriction of U* to M, viewed as an operator in M, and by U_o^* the adjoint of U_o. Then, for $f,g \in M$:

$$(f,U_o g) = (f,Ug) = (U^*f,g) = ((U^*)_o f,g),$$

hence $U_o^* = (U^*)_o$. This implies that $U_o^* U_o = (U^*)_o U_o$. But for $f \in M$, $(U^*)_o U_o f = U^*Uf = f$, so that $U_o^* U_o = I_o$, the identity operator on M. Similarly $U_o U_o^* = I_o$, proving the unitarity of U_o. □

Problem 1.19 : (a) Let $\{e_i\}_{i=1}^{\infty}$ be an orthonormal basis of an infinite-dimensional Hilbert space. Define a linear operator Ω by its action on each e_i as follows : $\Omega e_i = e_{i+1}$. (α) Find Ω^*. (β) Show that Ω is isometric but not unitary, and determine $F = \Omega\Omega^*$.

(b) In a finite-dimensional Hilbert space, each isometric operator is unitary.

A generalization of the notion of an isometry is that of a partial isometry. An operator Ω in $B(H)$ is called a partial isometry if

$$\Omega^*\Omega = E, \tag{1.47}$$

where E is a projection. Some properties of partial isometries are given in the next proposition.

Proposition 1.20 : Let Ω be a partial isometry. Then

(a) $\Omega E = \Omega$, (1.48)

(b) $(\Omega f, \Omega g) = (Ef, Eg) \quad \forall\, f, g \in H$, (1.49)

(c) $\|\Omega\| = 1$ unless $E = 0$,

(d) $\Omega\Omega^*$ is a projection, and $M(\Omega\Omega^*) = R(\Omega)$.

Proof : Let $f \in H$. Then, by (1.47),

$$\|\Omega Ef - \Omega f\|^2 = (\Omega(Ef - f), \Omega(Ef - f)) = (Ef - f, \Omega^*\Omega(Ef - f))$$

$$= (Ef - f, E(Ef - f) = 0,$$

since $E(Ef - f) = E^2 f - Ef = Ef - Ef = \theta$. Hence $\Omega Ef - \Omega f = \theta$.

The proof of (b)-(d) is very similar to that of the corresponding statements in Proposition 1.17 and left as an exercise. □

1.3 Integration in Hilbert Space

A vector-valued function f on an interval $\Delta \subseteq \mathbb{R}$ is a mapping from Δ into a Hilbert space H. Such a function associates with each point $s \in \Delta$ a vector $f(s)$ in H. f is called strongly continuous if, for each $t \in \Delta$,

$$\lim_{s \to t, s \in \Delta} \|f(s) - f(t)\| = 0. \tag{1.50}$$

Similarly, an operator-valued function is a mapping A from Δ into $B(H)$ and is called strongly continuous if, for each $t \in \Delta$,

$$\underset{s\to t, s\in\Delta}{\text{s-lim}}\ A(s) = A(t). \tag{1.51}$$

As in the case of real- or complex-valued functions one can define the Riemann integral for vector-valued and operator-valued functions. The proofs of the properties of these integrals are obtained by mimicking those for the scalar-valued integral, replacing the absolute value $|f(s)|$ by the norm $\|f(s)\|$ or $\|A(s)\|$. We do not give these proofs here. They may be found in Chapter 4-4 of [AJS].

First let $\Delta = [a,b]$ be a finite interval. A <u>partition</u> Π of $[a,b]$ is a set of numbers $\{s_0, s_1, \ldots, s_n; u_1, \ldots, u_n\}$ such that $a = s_0 < s_1 < \ldots < s_n = b$ and $u_k \in (s_{k-1}, s_k]$. We set $|\Pi| = \max_{k=1,\ldots,n} |s_k - s_{k-1}|$.

Given a vector-valued function $f:[a,b] \to H$, we set

$$\Sigma(\Pi,f) = \sum_{k=1}^{n} f(u_k)(s_k - s_{k-1}). \tag{1.52}$$

(1.52) is a finite sum of vectors in H and thus gives a well defined vector $\Sigma(\Pi,f)$ in H. One now takes a sequence $\{\Pi_r\}$ of partitions such that $|\Pi_r| \to 0$ as $r \to \infty$ and defines

$$\int_a^b f(s)ds = \underset{r\to\infty}{\text{s-lim}}\ \Sigma(\Pi_r, f) \tag{1.53}$$

if the limit exists and is independent of the sequence $\{\Pi_r\}$.

If Δ is an infinite interval, one first defines the integral on each finite subinterval as in (1.53) and then lets the length of the subinterval tend to infinity in an appropriate manner. Thus for example, if a is finite and $b = \infty$, one defines (if the limit exists)

$$\int_a^\infty f(s)ds = \underset{b\to\infty}{\text{s-lim}} \int_a^b f(s)ds. \tag{1.54}$$

Proposition 1.21 : Let [a,b] and [b,c] be finite or infinite intervals, and assume that all integrals below exist. Then

(a) $$\int_a^b f(s)ds + \int_b^c f(s)ds = \int_a^c f(s)ds. \tag{1.55}$$

(b) $$\int_a^b\{\alpha f_1(s) + f_2(s)\}ds = \alpha\int_a^b f_1(s)ds + \int_a^b f_2(s)ds. \tag{1.56}$$

(c) $$\left\|\int_a^b f(s)ds\right\| \le \int_a^b\|f(s)\|ds. \tag{1.57}$$

(d) If $A\in\mathcal{B}(H)$, then

$$A\int_a^b f(s)ds = \int_a^b Af(s)ds. \tag{1.58}$$

Proof of (c) : (i) Assume a and b are finite. Then (1.57) is obtained by using Proposition 1.1, the triangle inequality (1.11) and the definition of the scalar-valued Riemann integral :

$$\left\|\int_a^b f(s)ds\right\| = \lim_{|\Pi|\to 0}\|\Sigma(\Pi,f)\|$$

$$\le \lim_{|\Pi|\to 0}\sum_{k=1}^{n}\|f(u_k)\|\,|s_k - s_{k-1}| = \int_a^b\|f(s)\|ds.$$

(ii) If for example a is finite and b = ∞, one obtains from Proposition 1.1 and (i) above that

$$\left\|\int_a^\infty f(s)ds\right\| = \lim_{b\to\infty}\left\|\int_a^b f(s)ds\right\| \le \lim_{b\to\infty}\int_a^b\|f(s)\|ds$$

$$= \int_a^\infty\|f(s)\|ds. \qquad \square$$

Definition : A vector-valued function $f:\Delta\to H$ is strongly differentiable if there is a function $f':\Delta\to H$ such that

$$\lim_{\tau\to 0}\left\|\frac{1}{\tau}[f(s+\tau) - f(s)] - f'(s)\right\| = 0 \quad \text{for each } s\in\Delta.$$

We also write

$$f'(s) = \frac{d}{ds}f(s) = \text{s-}\lim_{\tau\to 0} \tau^{-1}[f(s+\tau) - f(s)]. \quad (1.59)$$

Similarly one defines the <u>strong derivative</u> A' of an operator-valued function $A:\Delta \to \mathcal{B}(\mathcal{H})$ as

$$A'(s) = \text{s-}\lim_{\tau\to 0} \tau^{-1}[A(s+\tau) - A(s)]. \quad (1.60)$$

<u>Proposition 1.22</u> :

(a) If [a,b] is finite and $f:[a,b] \to \mathcal{H}$ strongly continuous, then $\int_a^b f(s)ds$ exists.

(b) If a,b are arbitrary, f is strongly continuous on (a,b) and $\int_a^b \|f(s)\| ds < \infty$, then $\int_a^b f(s)ds$ exists.

(c) If f is strongly differentiable on (a,b) and f' is strongly continuous and integrable, then

$$\int_a^b f'(s)ds = f(b) - f(a). \quad (1.61)$$

<u>Problem 1.23</u> : Prove part (b) of Proposition 1.22 from part (a).

We now turn to the integral of operator-valued functions. If $A:\Delta \to \mathcal{B}(\mathcal{H})$ is such a function, then the correspondence $s \mapsto A(s)f$ defines, for each fixed f in $\mathcal{H}$, a vector-valued function. If, for each $f \in \mathcal{H}$, the latter is integrable over Δ, we may define an operator $\int_\Delta A(s)ds$ with domain $\mathcal{H}$ by

$$[\int_\Delta A(s)ds]f = \int_\Delta A(s)f\,ds . \quad (1.62)$$

The properties of the operator-valued integral follow immediately from the preceding two propositions. In particular :

Proposition 1.24 :

(a) $||\int_a^b A(s)ds|| \le \int_a^b ||A(s)||ds.$ (1.63)

(b) If $B \in \mathcal{B}(\mathcal{H})$, then

$$B \int_a^b A(s)ds = \int_a^b BA(s)ds. \qquad (1.64)$$

(c) If $[a,b]$ is a finite closed interval and $A:[a,b] \to \mathcal{B}(\mathcal{H})$ is strongly continuous, then $\int_a^b A(s)ds$ exists.

(d) If a,b are arbitrary, A is strongly continuous and $\int_a^b ||A(s)||ds < \infty$, then $\int_a^b A(s)ds$ exists.

Remark 1.25 : In Proposition 1.24 (c), the proof of the existence of $\int_a^b A(s)ds$ involves only the norm in $\mathcal{H}$ but not the scalar product. One shows that, if the integrand is continuous, the sequence $\{\Sigma(\Pi_r,g)\}$ is strongly Cauchy for each $f \in \mathcal{H}$, where $g:\Delta \to \mathcal{H}$ is defined by $g(s) = A(s)f$. Since $\mathcal{H}$ is complete, this sequence has a limit in $\mathcal{H}$.

If one assumes A(s) to be continuous in a different norm (e.g. in the operator-norm, i.e. $||A(s) - A(t)|| \to 0$ as $s \to t$), then the sums $\sum_{k=1}^n A(u_k)(s_k - s_{k-1})$ will be Cauchy in this other norm. If all operators A(s), $s \in [a,b]$, belong to a subset $\mathcal{A}$ of $\mathcal{B}(\mathcal{H})$ which is complete with respect to this norm, the integral $\int_a^b A(s)ds$ will also be in $\mathcal{A}$. This observation will be used later, in particular for $\mathcal{A} = \mathcal{B}_\infty$ (the set of compact operators introduced in Chapter 3).

CHAPTER 2 : SELF-ADJOINT OPERATORS. SCHRÖDINGER OPERATORS

In Section 2.1 we define symmetric and self-adjoint operators and give criteria for a symmetric operator to be self-adjoint. In Section 2.2 we study simple spectral properties of self-adjoint operators. A particular class of self-adjoint operators, the so-called multiplication operators, are introduced in Section 2.3, and the results are applied to proving the essential self-adjointness of the Laplacian. In Section 2.4 we give a criterion for the invariance of self-adjointness under perturbations and apply it to Schrödinger operators with non-singular potentials. Finally, in Section 2.5, we give a characterization of the domain of Schrödinger operators with strongly singular potentials. The importance of self-adjointness will be discussed in Section 4.1.

2.1 Self-adjointness Criteria

Throughout this chapter we assume that A is a linear operator with dense domain $D(A)$, so that A^* is defined.

Definition : A is symmetric if $D(A)$ is dense in H and $A \subseteq A^*$ (i.e. if $D(A) \subseteq D(A^*)$ and $A^*f = Af$ for each $f \in D(A)$).

The condition $A \subseteq A^*$ may also be written as

$$(Af,g) = (f,Ag) \quad \text{for all} \quad f,g \in D(A). \tag{2.1}$$

Definition : A is self-adjoint if $D(A)$ is dense in H and $A^* = A$ (i.e. if $D(A) = D(A^*)$ and $A^*f = Af$ for each $f \in D(A)$).

Clearly each self-adjoint operator is symmetric. If A is bounded and $D(A) = H$, then A is symmetric if and only if it is self-adjoint (this follows immediately from the definitions). If A is unbounded, the condition that A be self-adjoint is a very strong condition, since it requires that D(A*) be exactly the same as D(A). The condition (2.1), which is often easy to verify in applications, is not sufficient for A to be self-adjoint.

If A is a symmetric operator, one can try to extend it to a linear operator A_1 such that $D(A_1)$ is strictly larger than D(A). It follows from the definition (1.32) of the domain of the adjoint that $D(A_1^*)$ is smaller than (or possibly equal to) D(A*), since a vector in $D(A_1^*)$ has to satisfy more conditions than a vector in D(A*). We can imagine that in this way it may be possible to find extensions A_1 of A such that $D(A_1^*) = D(A_1)$, in other words self-adjoint extensions of A.

The theory of symmetric extensions of a symmetric operator is an important topic in functional analysis, which will however not be treated in these lectures. What we shall do is to prove that certain symmetric operators are self-adjoint or essentially self-adjoint (cf. the definition below). We wish to point out, though, that there are symmetric operators having no self-adjoint extension at all and also symmetric operators having an uncountable number of self-adjoint extensions. Examples are easily found among ordinary differential operators (see e.g. [AG, no. 49]).

A self-adjoint operator clearly has no proper self-adjoint extension, since any proper extension is not symmetric any more. Thus a self-adjoint operator A has exactly one self-adjoint extension (an improper one), namely A itself. More generally, a symmetric operator which has one and only one (proper or improper) self-adjoint extension is said to be essentially self-adjoint, since it determines a unique self-adjoint operator.

If A is an arbitrary symmetric operator, then A^* has dense domain, hence the operator A^{**} is also defined, and one has $A \subseteq A^{**}$ by (1.37). Furthermore the symmetry of A, i.e. the relation $A \subseteq A^*$, implies that $A^{**} \subseteq A^*$ (see Problem 1.12 (e)). Since $A^* = A^{***}$, this shows that A^{**} is symmetric. Hence A^{**} is always a symmetric extension of A. The first lemma shows that, if A^{**} is self-adjoint, then it is the only self-adjoint extension of A.

Lemma 2.1 : If $A \subseteq A^*$ and $A^{**} = A^*$, then A is essentially self-adjoint (and $A^* = A^{**}$ is its unique self-adjoint extension).

Proof : From the preceding remarks and the hypothesis $A^{**} = A^*$, we know that A has at least one self-adjoint extension, namely A^*. So assume that B is an arbitrary self-adjoint extension of A, i.e. $A \subseteq B$ and $B^* = B$. By Problem 1.12 (e) we then have $B = B^* \subseteq A^*$, hence (applying again Problem 1.12 (e)) $A^{**} \subseteq B^* = B$. By combining these relations with $A^{**} = A^*$, we obtain that $A^* = A^{**} \subseteq B \subseteq A^*$. Hence we must have equality, which shows that $B = A^*$. □

We now give the basic self-adjointness criterion.

Proposition 2.2 : Let A be a symmetric operator. Then A is self-adjoint if and only if $R(A + iI) = H$ and $R(A - iI) = H$.

Proof : (i) Suppose $R(A \pm iI) = H$. Since $A \subseteq A^*$, we can conclude that $A = A^*$ as soon as we know that

$$D(A^*) \subseteq D(A). \tag{2.2}$$

For this, let $g \in D(A^*)$. Since $R(A - iI) = H$, there is a vector $h \in D(A)$ such that $(A^* - i)g = (A - i)h$. Now $(A - i)h = (A^* - i)h$ by the symmetry of A, hence $(A^* - i)(g - h) = \theta$. Since $R(A + iI) = H$, we have $N((A + i)^*) = N(A^* - i) = \{\theta\}$, see Proposition 1.13. Therefore $g - h = \theta$, hence $g = h \in D(A)$, which proves (2.2).

(ii) Suppose $A = A^*$. Then one has for each $f \in D(A)$

$$\|(A \pm i)f\|^2 = \|Af\|^2 + \|f\|^2 \pm i(Af,f) \mp i(f,Af)$$
$$= \|Af\|^2 + \|f\|^2. \qquad (2.3)$$

We use this to show that $R(A \pm iI)$ are subspaces. For this, let $\{g_n\}$ be a Cauchy sequence in $R(A+i)$, say. Thus $g_n = (A+i)f_n$, with $f_n \in D(A)$, and $s\text{-}\lim g_n = g$. We must show that $g = (A+i)f$ for some $f \in D(A)$.

Now by (2.3)

$$\|g_n - g_m\|^2 = \|Af_n - Af_m\|^2 + \|f_n - f_m\|^2. \qquad (2.4)$$

Since $\{g_n\}$ is Cauchy, (2.4) implies that $\{f_n\}$ is also a Cauchy sequence. We set $f = s\text{-}\lim f_n$. Since $A = A^*$, we have $A + i = (A-i)^*$, so that $A + i$ is closed by Proposition 1.14. It follows from the definition of closedness that $f \in D(A+i) = D(A)$ and that $(A+i)f = s\text{-}\lim(A+i)f_n = g$. This proves that $R(A+iI) = H$. Similarly we find that $R(A-iI) = H$.

It remains to show that the subspaces $R(A \pm i)$ are equal to H, in other words that $h \perp R(A \pm i)$ implies that $h = \theta$. Now if $h \perp R(A+i)$, then $h \in N(A^* - i) = N(A-i)$, so that $Ah = ih$. Therefore

$$(h,h) = (h,-iAh) = (iA^*h,h) = (iAh,h) = -(h,h),$$

so that $(h,h) = 0$, whence $h = \theta$. Similarly, if $h \perp R(A-i)$, then $h = \theta$. □

<u>Proposition 2.3</u> : Suppose $A \subseteq A^*$ and $R(A + \mu I) = H$ for some real number μ. Then A is self-adjoint.

<u>Proof</u> : The proof is essentially the same as part (i) of the pre-

ceding proof. It suffices to replace $\pm i$ by μ. □

Definition : A linear operator is positive, written $A \geq 0$, if $(f,Af) \geq 0$ for all $f \in D(A)$.

Proposition 2.4 : Suppose A is symmetric and positive. Then, if $R(A+I)$ is dense in H, A is essentially self-adjoint, and its unique self-adjoint extension is A^*.

Proof : Since $A \geq 0$, we have for each $f \in D(A)$

$$\|(A+I)f\|^2 = \|Af\|^2 + \|f\|^2 + 2(f,Af) \geq \|Af\|^2 + \|f\|^2. \quad (2.5)$$

(i) We first show that $R(A^{**}+I) = H$. For this, let $g \in H$ and choose a sequence $\{f_n\} \in D(A)$ such that $g = \text{s-lim}(A+I)f_n$. Since $\{(A+I)f_n\}$ is Cauchy, we obtain from (2.5) that $\{f_n\}$ is Cauchy (use the argument given in relation to (2.4)). We set $f = \text{s-lim}\, f_n$.

Since $A \subseteq A^{**}$, we have $f_n \in D(A^{**})$, $\text{s-lim}\, f_n = f$ and $\text{s-lim}(A^{**}+I)f_n = g$. Now $A^{**}+I$ is closed (see Proposition 1.14), so that, by the definition of closedness, $f \in D(A^{**}+I) = D(A^{**})$ and $(A^{**}+I)f = g$. This verifies our claim that $R(A^{**}+I) = H$.

(ii) We now show that $A^* = A^{**}$. Since $A \subseteq A^*$, we have $A^{**} \subseteq A^*$, and it suffices to show that $D(A^*) \subseteq D(A^{**})$. So let $f \in D(A^*)$. Then, by (i), there is a vector $g \in D(A^{**})$ such that $(A^*+I)f = (A^{**}+I)g = (A^*+I)g$. Thus $(A^*+I)(f-g) = \theta$. Since $N(A^*+I) = R(A+I)^{\perp} = \{\theta\}$, we have $f = g \in D(A^{**})$.

(iii) The result of the proposition now follows from that of (ii) and Lemma 2.1. □

2.2 Spectral Properties of Self-Adjoint Operators

Definition : A complex number λ is an eigenvalue of the linear operator B if there exists a vector $f \neq \theta$ in $D(B)$ such that $Bf = \lambda f$.

f is called an eigenvector of B associated with λ.

The set of all eigenvectors associated with a complex number λ is nothing but the null space $N(B - \lambda)$ of $B - \lambda$. If B is closed, this set is a subspace of H called the eigensubspace M_λ associated with λ (cf. Problem 1.15).

Proposition 2.5 : Let A be self-adjoint. Then

(a) Each eigenvalue of A is real,

(b) The eigensubspaces corresponding to different eigenvalues are mutually orthogonal : if $Af_1 = \lambda_1 f_1$, $Af_2 = \lambda_2 f_2$ and $\lambda_1 \neq \lambda_2$, then $f_1 \perp f_2$.

Proof : (a) Assume $Af = \lambda f$, $\lambda \in \mathbb{C}$. Then

$$\lambda(f,f) = (f,\lambda f) = (f,Af) = (Af,f) = (\lambda f,f) = \bar{\lambda}(f,f).$$

Since $f \neq \theta$, we must have $\lambda = \bar{\lambda}$, hence $\lambda \in \mathbb{R}$.

(b) Since λ_i are real, we get that

$$(\lambda_1 - \lambda_2)(f_1,f_2) = (\lambda_1 f_1,f_2) - (f_1,\lambda_2 f_2) = (Af_1,f_2) - (f_1,Af_2) = 0.$$

Since $\lambda_1 \neq \lambda_2$, this implies that $(f_1,f_2) = 0$. □

Corollary 2.6 : A self-adjoint operator in a separable Hilbert space has at most countably many eigenvalues.

Proof : Let $\{f_\alpha\}$ be a collection of eigenvectors belonging to different eigenvalues λ_α, and $\|f_\alpha\| = 1$. Then $\{f_\alpha\}$ forms an orthonormal set in H. If H is separable, each orthonormal set contains at most countably many elements. □

Each self-adjoint operator induces naturally a decomposition of the underlying Hilbert space H into a direct sum of two orthogonal subspaces. This is obtained as follows. Define $H_p(A)$ to be the subspace spanned by all eigenvectors of A, i.e. the closure of the linear manifold of all finite linear combinations of eigenvectors of A. Alternatively, $H_p(A)$ is the direct sum of all eigensubspaces of A : $H_p(A) = \oplus M_i = \oplus N(A - \lambda_i)$, where $\{\lambda_i\}$ are the eigenvalues of A.

If we now define $H_c(A)$ to be the orthogonal complement of $H_p(A)$, we see that H is the direct sum of $H_p(A)$ and $H_c(A)$:

$$H = H_p(A) \oplus H_c(A). \qquad (2.6)$$

Thus each vector f in H has a unique decomposition as

$$f = f_p \oplus f_c \qquad (2.7)$$

with $f_p \in H_p(A)$, $f_c \in H_c(A)$ and $(f_p, f_c) = 0$.

The indices p and c stand for "point spectrum" and "continuous spectrum". Indeed we shall show below that the restrictions of A to $H_p(A)$ and $H_c(A)$ define self-adjoint operators A_p and A_c in these subspaces; A_p has a complete set of eigenvectors (the eigenvectors of A), in other words pure point spectrum. The operator A_c has no eigenvalues; it is an operator with purely continuous spectrum. $H_c(A)$ is therefore called the <u>subspace of continuity</u> of A.

If $H_p(A) = H$, $H_c(A) = \{\theta\}$, then A is said to have <u>pure point spectrum</u>. An example is the Hamiltonian of the harmonic oscillator $A = \underline{P}^2 + \underline{Q}^2$ in $L^2(\mathbb{R}^\nu)$. If, on the contrary, $H_p(A) = \{\theta\}$ and $H_c(A) = H$, A is said to have <u>purely continuous spectrum</u>. An example is the free Hamiltonien $H_o = \underline{P}^2$ of non-relativistic quantum mechanics which will be introduced in the next section.

<u>Proposition 2.7</u> : Let A be a self-adjoint operator in a Hilbert

space $\mathcal{H}$. Let A_p and A_c be the restrictions of A to $D(A) \cap \mathcal{H}_p(A)$ and $D(A) \cap \mathcal{H}_c(A)$, respectively. Then A_p leaves $\mathcal{H}_p(A)$ invariant and A_c leaves $\mathcal{H}_c(A)$ invariant. One may therefore view A_p to be an operator in $\mathcal{H}_p(A)$ and A_c to be an operator in $\mathcal{H}_c(A)$. With this convention, A_p and A_c are self-adjoint operators in $\mathcal{H}_p(A)$ and $\mathcal{H}_c(A)$ respectively, and one may write, in the decomposition (2.6) of $\mathcal{H}$:

$$A = A_p \oplus A_c. \tag{2.8}$$

Proof : Let $\{e_r\}$ be an orthonormal basis of $\mathcal{H}_p(A)$ formed of eigenvectors of A, $Ae_r = \lambda_r e_r$. Then, for each $f \in \mathcal{H}$,

$$f_p = \sum_r (e_r, f) e_r. \tag{2.9}$$

(i) Let $f, g \in D(A)$. Then, by (2.9),

$$\begin{aligned}(f_p, Ag) &= \sum_r ((e_r, f)e_r, Ag) = \sum_r \overline{(e_r, f)}(e_r, Ag) \\ &= \sum_r \overline{(e_r, f)}\lambda_r (e_r, g) = \sum_r \overline{(Ae_r, f)}(e_r, g) \\ &= \sum_r \overline{(e_r, Af)}(e_r, g) = \sum_r ((e_r, Af)e_r, g) = ((Af)_p, g).\end{aligned}$$

This shows that $f_p \in D(A^*) = D(A)$ and that

$$Af_p = A^* f_p = (Af)_p. \tag{2.10}$$

(ii) Taking in (i) $f = f_p \in D(A) \cap \mathcal{H}_p(A)$, we see from (2.10) that A leaves $\mathcal{H}_p(A)$ invariant. On the other hand, if $h \in \mathcal{H}_c(A) \cap D(A)$, one has for each eigenvector f_i of A ($Af_i = \lambda_i f_i$)

$$(Ah, f_i) = (h, Af_i) = \lambda_i (h, f_i) = 0$$

since $h \perp f_i$. Thus $Ah \perp \mathcal{H}_p(A)$, i.e. A leaves $\mathcal{H}_c(A)$ invariant.

(iii) Let $f \in \mathcal{H}_p(A)$ and $\varepsilon > 0$. There is a $g \in D(A)$ such that

$\|f - g\| < \varepsilon$, in other words $\|f - g\|^2 = \|f - g_p\|^2 + \|g_c\|^2 < \varepsilon^2$. Hence $\|f - g_p\| < \varepsilon$. Since $g_p \in D(A_p)$ by (i), this shows that $D(A_p)$ is dense in $H_p(A)$.

Finally let A_p^* be the adjoint of A_p, as an operator in $H_p(A)$. Let $f \in D(A_p^*) \subseteq H_p(A)$ and $g \in D(A)$. Then, by (2.10),

$$(f,Ag) = (f,(Ag)_p + (Ag)_c) = (f,(Ag)_p) = (f,Ag_p)$$

$$= (f,A_p g_p) = (A_p^* f,g_p) = (A_p^* f,g_p + g_c) = (A_p^* f,g).$$

In view of the definition of the adjoint operator, this means that $f \in D(A^*)$ and that $A^*f = A_p^*f$. Since $A^* = A$, this in turn means that $A_p^*f = Af = A_p f$, so that $A_p^* = A_p$. Similarly one shows that $A_c^* = A_c$. □

<u>Definition</u> : Let A be a closed linear operator. The complex number z is called a <u>regular point</u> of A if

i) $(A - zI)$ is invertible,

ii) $D((A - zI)^{-1}) = H$,

iii) $(A - zI)^{-1}$ is bounded,

in other words if $(A - zI)^{-1}$ exists and is in $B(H)$. The set of all regular points is called the <u>resolvent set</u> of A and denoted by $\rho(A)$. The complement $\sigma(A)$ of $\rho(A)$ in $\mathbb{C}$ is called the <u>spectrum</u> of A :

$$\sigma(A) := \mathbb{C}\backslash\rho(A). \tag{2.11}$$

The spectrum of the operator A_p is called the <u>point spectrum</u> $\sigma_p(A)$ of A and the spectrum of A_c the <u>continuous spectrum</u> $\sigma_c(A)$ of A. Thus, by definition,

$$\sigma_p(A) = \sigma(A_p), \qquad \sigma_c(A) = \sigma(A_c). \tag{2.12}$$

One can show that $\rho(A)$ is an open subset of $\mathbb{C}$, hence the spectrum $\sigma(A)$ is closed. If A is self-adjoint, its spectrum lies entirely on the real axis :

<u>Proposition 2.8</u> : Let $A = A^*$ and $z = x + iy$ with $x,y \in \mathbb{R}$ and $y \neq 0$. Then $z \in \rho(A)$.

<u>Proof</u> : (i) As in (2.3), we have for $f \in D(A)$:

$$\|(A - z)f\|^2 = \|(A - x)f\|^2 + y^2\|f\|^2. \tag{2.13}$$

Now suppose that $(A - z)f = \theta$. Then (2.13) implies that $f = \theta$, since $y \neq 0$. Hence $A - zI$ is invertible. Also, if we set in (2.13) $f = (A - z)^{-1}g$, $g \in R(A - z)$, we obtain

$$\|g\|^2 \geq y^2\|(A - z)^{-1}g\|^2.$$

Hence $(A - zI)^{-1}$ is bounded, and

$$\|(A - zI)^{-1}\| \leq |y|^{-1}. \tag{2.14}$$

(ii) As x and y are real and $y \neq 0$, the operator $y^{-1}(A - x)$ is self-adjoint (cf. Problem 1.12 (d)). By Proposition 2.2, $R(y^{-1}(A - x) - i) = H$. Thus, for each $g \in H$ there is an $f \in D(A)$ such that

$$\frac{1}{y}g = \left[\frac{1}{y}(A - x) - i\right]f,$$

or $g = (A - x - iy)f = (A - z)f$. This shows that $R(A - zI) \equiv$ $\equiv D((A - zI)^{-1}) = H$. □

If λ is an eigenvalue of A, $N(A - \lambda)$ is a non-empty subspace of H, hence $A - \lambda I$ is not invertible. If λ is not an eigenvalue but belongs to the continuous spectrum of A, then $A - \lambda I$ is invertible but $D((A - \lambda I)^{-1})$ is only a dense subset of H different from H. An example will be given in Proposition 2.17 (c) (see part (iii) of its proof).

Definition : Let A be a closed operator. Then the operator-valued function $z \mapsto (A - zI)^{-1}$ from $\rho(A)$ to $\mathcal{B}(\mathcal{H})$ is called the resolvent of A. The value of the resolvent at a point $z \in \rho(A)$, i.e. the operator $(A - zI)^{-1}$ for fixed $z \in \rho(A)$, is sometimes also simply called the resolvent. (We shall denote the resolvent indifferently by $(A - zI)^{-1}$ or by $(A - z)^{-1}$.)

Proposition 2.9 : Let A be a closed operator and $z, z_1, z_2 \in \rho(A)$. Then

(a) $(A - zI)^{-1}$ maps $\mathcal{H}$ onto D(A) and

$$A(A - zI)^{-1}f = (A - zI)^{-1}Af \qquad \forall\, f \in D(A).$$

(b) The following identity, called the first resolvent equation, holds :

$$(A - z_1 I)^{-1} - (A - z_2 I)^{-1} = (z_1 - z_2)(A - z_1)^{-1}(A - z_2)^{-1}. \qquad (2.15)$$

(c) $(A - z_1 I)^{-1}(A - z_2 I)^{-1} = (A - z_2 I)^{-1}(A - z_1 I)^{-1}$, $\qquad$ (2.16)

i.e. the resolvent at a point $z_1 \in \rho(A)$ commutes with the resolvent at any other point $z_2 \in \rho(A)$.

Proof : (a) One has from the definition of the inverse operator that $\mathcal{R}((A - zI)^{-1}) = D(A - zI) = D(A)$. If $f \in D(A)$, then

$$(A - z)^{-1}Af = (A - z)^{-1}(A - z)f + z(A - z)^{-1}f = f + z(A - z)^{-1}f$$

$$= (A - z)(A - z)^{-1}f + z(A - z)^{-1}f = A(A - z)^{-1}f.$$

(b) One formally has

$$(z_1 - z_2)(A - z_1)^{-1}(A - z_2)^{-1} = (A - z_1)^{-1}[(A - z_2) - (A - z_1)](A - z_2)^{-1}$$

$$= (A - z_1)^{-1} - (A - z_2)^{-1}.$$

The first equality makes sense since $R((A-z_2)^{-1}) = D(A)$, so that $(A-z_i)(A-z_2)^{-1}f$ is defined for each $f \in H$. The second equality holds since $(A-z_2)(A-z_2)^{-1}f = f$ for each $f \in H$.

(c) If $z_1 = z_2$, (2.16) is trivial. If $z_1 \neq z_2$, it follows immediately from (2.15). □

Proposition 2.10 : Let A be a closed operator. Then

(a) The mapping $z \mapsto (A-zI)^{-1}$ is continuous in the operator norm on $\rho(A)$, i.e.

$$\underset{\substack{z_1 \to z \\ z, z_1 \in \rho(A)}}{\text{u-lim}} \|(A-z)^{-1} - (A-z_1)^{-1}\| = 0. \tag{2.17}$$

(b) The resolvent is differentiable in operator norm, and

$$\frac{d}{dz}(A-zI)^{-1} := \underset{z_1 \to z}{\text{u-lim}}(z_1 - z)^{-1}[(A-z_1)^{-1} - (A-z)^{-1}] = (A-zI)^{-2}. \tag{2.18}$$

Proof : From (2.15),

$$(A-z_1)^{-1} = (A-z)^{-1} + (z_1 - z)(A-z_1)^{-1}(A-z)^{-1}. \tag{2.19}$$

By inserting this expression for $(A-z_1)^{-1}$ into the r.h.s. of (2.19) and iterating the procedure, one obtains that

$$(A-z_1)^{-1} - (A-z)^{-1} = (z_1 - z)(A-z)^{-2} \sum_{k=0}^{n} [(z_1 - z)(A-z)^{-1}]^k$$
$$+ (z_1 - z)^{n+2}(A-z_1)^{-1}(A-z)^{-(n+2)}. \tag{2.20}$$

If $|z_1 - z| < \|(A-z)^{-1}\|^{-1}$, the second term on the r.h.s converges to zero in operator norm as $n \to \infty$, and the series in the first term is convergent in the operator norm (cf. Proposition 1.11). Hence

$$(A - z_1)^{-1} - (A - z)^{-1} = (z_1 - z)(A - z)^{-2} \sum_{k=0}^{\infty} [(z_1 - z)(A - z)^{-1}]^k,$$

and by Proposition 1.11

$$\|(A - z_1)^{-1} - (A - z)^{-1}\| \le |z_1 - z| \, \|(A - z)^{-1}\|^2 (1 - |z_1 - z| \, \|(A - z)^{-1}\|)^{-1},$$

which converges to zero as $z_1 \to z$. This proves (2.17). Now (2.18) follows very easily from (2.19) and (2.17). □

2.3 Multiplication Operators. The Laplacian

In this section we look at a simple class of self-adjoint operators, namely multiplication operators by real-valued functions in L^2-spaces. We shall use these results only for the case where the underlying Hilbert space is $L^2(\Delta)$, where Δ is an open subset of $\mathbb{R}^\nu$. We shall therefore treat only this case, although it would be easy to apply the theory to more general L^2-spaces.

Definition : Let Δ be an open set in $\mathbb{R}^\nu$ and $a : \Delta \to \mathbb{C}$ a measurable function. The multiplication operator A associated with a is the following linear operator in $L^2(\Delta)$:

$$D(A) = \{f \in L^2(\Delta) \mid \int_\Delta |a(\underline{x})|^2 |f(\underline{x})|^2 d^\nu x < \infty\}, \tag{2.21}$$

and $(Af)(\underline{x}) = a(\underline{x}) f(\underline{x})$ for $f \in D(A)$. (2.22)

Clearly D(A) is the maximal domain on which multiplication by $a(\underline{x})$ makes sense.

Proposition 2.11 : Let $a : \Delta \to \mathbb{R}$ be measurable and $|a(\underline{x})| < \infty$ a.e. Then the associated multiplication operator A is a self-adjoint operator in $L^2(\Delta)$.

Proof : (i) The inequality (1.12) implies that D(A) is a linear manifold in $\mathcal{H}$. To check that D(A) is dense in $\mathcal{H}$, it suffices to show that $f \perp D(A)$ implies $f = \theta$ (cf. Lemma 1.5).

For $m = 1,2,\ldots$, define a measurable subset Δ_m of Δ by $\Delta_m = \{\underline{x}\in\Delta \,|\, |a(\underline{x})| \le m\}$. The subspace $L^2(\Delta_m)$ of $L^2(\Delta)$ is contained in $D(A)$; in fact, if $f\in L^2(\Delta)$ is such that $f(\underline{x}) = 0$ for $\underline{x}\notin\Delta_m$, then

$$\int_\Delta |a(\underline{x})|^2 |f(\underline{x})|^2 d^\nu x = \int_{\Delta_m} |a(\underline{x})|^2 |f(\underline{x})|^2 d^\nu x$$

$$\le m^2 \int_{\Delta_m} |f(\underline{x})|^2 d^\nu x = m^2 \|f\|^2 < \infty. \qquad (2.23)$$

Hence, if $f \perp D(A)$, one has $f \perp L^2(\Delta_m)$ for each m. Thus, by Lemma 1.5, $f(\underline{x}) = 0$ on Δ_m, for each $m = 1,2,\ldots$. Since $|a(\underline{x})| < \infty$ a.e., the complement in Δ of $\cup_m \Delta_m$ is a null set (the set where $|a(\underline{x})| = \infty$). Hence $f(\underline{x}) = 0$ for all $\underline{x}\in\Delta$ except for $\underline{x}$ in the null set $\Delta \setminus \cup_m \Delta_m$, i.e. f is the zero vector in $L^2(\Delta)$.

(ii) Since a is real-valued, A is symmetric :

$$(f,Ag) = \int_\Delta \overline{f(\underline{x})} a(\underline{x}) g(\underline{x}) d^\nu x = (Af,g) \qquad \forall\, f,g\in D(A).$$

(iii) For $f\in L^2(\Delta)$, define $f_\pm$ by $f_\pm(\underline{x}) = [a(\underline{x}) \pm i]^{-1} f(\underline{x})$. Since $a(\underline{x})$ is real, one has

$$|a(\underline{x}) \pm i|^{-1} \le 1 \quad \text{and} \quad |a(\underline{x})|\,|a(\underline{x}) \pm i|^{-1} \le 1.$$

This implies as in (2.23) that $\|f_\pm\|^2 \le \|f\|^2$, hence $f_\pm \in L^2(\Delta)$, and that $\|Af_\pm\|^2 \le \|f\|^2$, hence $f_\pm \in D(A)$. Clearly $(A \pm i)f_\pm = f$. Since f was an arbitrary vector in $L^2(\Delta)$, this means that $\mathcal{R}(A \pm iI) = \mathcal{H}$, so that $A = A^*$ by Proposition 2.2. □

The next proposition uses the definition of the L^∞-norm which is given in the Appendix.

Proposition 2.12 : Let A be the multiplication operator associated with a function $a : \Delta \to \mathbb{C}$. Then A is in $\mathcal{B}(L^2(\Delta))$ if and only if

$\|a\|_\infty < \infty$, in which case

$$\|A\| = \|a\|_\infty. \tag{2.24}$$

<u>Proof</u> : (i) Suppose $\|a\|_\infty < \infty$. Then one has as in (2.23) that $\|Af\|^2 \le \|a\|_\infty^2 \|f\|^2$ for each $f \in L^2(\Delta)$, hence $D(A) = L^2(\Delta)$ and $\|A\| \le \|a\|_\infty$.

Now let $N < \|a\|_\infty$ and set $\Delta(N) = \{\underline{x} \in \Delta \,\big|\, |a(\underline{x})| > N\}$. $\Delta(N)$ has positive measure (otherwise one would have $\|a\|_\infty \le N$). Therefore there exists a vector $f \neq \theta$ in $L^2(\Delta)$ such that $f(\underline{x}) = 0$ for $\underline{x} \notin \Delta(N)$. Now $\|Af\|^2 \ge N^2 \|f\|^2$, hence $\|A\| \ge N$. Thus $N \le \|A\| \le \|a\|_\infty$ for each $N < \|a\|_\infty$, which proves (2.24).

(ii) Assume $\|a\|_\infty = \infty$. For each $m > 0$, define $\Delta(m)$ as above. The measure of $\Delta(m)$ is positive, and as in (i) there is a vector $f \neq \theta$ in $L^2(\Delta)$ such that $\|Af\| \ge m\|f\|$. Since m is arbitrary, A cannot be bounded. □

We now give some examples of self-adjoint multiplication operators in $L^2(\mathbb{R}^\nu)$.

<u>Example 2.13</u> : We denote by $Q_m (m = 1,\ldots,\nu)$ the multiplication operator by x_m in $L^2(\mathbb{R}^\nu)$:

$$(Q_m f)(\underline{x}) = x_m f(\underline{x}). \tag{2.25}$$

It is called the m-th component of the <u>position operator</u> in quantum mechanics.

<u>Example 2.14</u> : We denote by $P_m (m = 1,\ldots,\nu)$ the multiplication operator by k_m in $\tilde{L}^2(\mathbb{R}^\nu)$:

$$(FP_m f)(\underline{k}) = k_m \tilde{f}(\underline{k}). \tag{2.26}$$

P_m is called the m-th component of the <u>momentum operator</u>.

Example 2.15 : We denote by H_0 the multiplication operator by $|\underline{k}|^2$ in $\tilde{L}^2(\mathbb{R}^\nu)$:

$$(FH_0 f)(\underline{k}) = |\underline{k}|^2 \tilde{f}(\underline{k}). \tag{2.27}$$

This operator is called the Schrödinger free Hamiltonian in quantum mechanics. It is related to the operators in Example 2.14 by

$$H_0 = \underline{P}^2 = \sum_{m=1}^{\nu} P_m^2. \tag{2.28}$$

Example 2.16 : If $v:\mathbb{R}^\nu \to \mathbb{R}$ is any measurable function which is finite a.e., it determines a multiplication operator V in $L^2(\mathbb{R}^\nu)$. We shall use the letters v and V for such operators when we have in mind the interaction operator of a non-relativistic quantum particle, and the function v will then be called a potential.

The operator H_0 is the fundamental operator in our study of non-relativistic quantum dynamics, and in the remainder of this section we give some further properties of this operator. The Hamiltonian for a particle moving under the influence of a potential v is formally given by $H = H_0 + V$, and the main topic of the next section will be to study this operator sum, and in particular to find conditions on v such that H is essentially self-adjoint.

Proposition 2.17 :

(a) H_0 is a positive unbounded operator. Its spectrum is $[0,\infty)$ and is purely continuous. In particular

$$\mathcal{H}_p(H_0) = \{\theta\}, \qquad \mathcal{H}_c(H_0) = \mathcal{H}. \tag{2.29}$$

(b) $D(H_0)$ is contained in $D(P_m)$, and $P_m(H_0 - zI)^{-1}$ belongs to $\mathcal{B}(\mathcal{H})$ for each complex $z \notin [0,\infty)$.

(c) The resolvent of H_0 is the multiplication operator in $\tilde{L}^2(\mathbb{R}^\nu)$

by $(\underline{k}^2 - z)^{-1}$.

Proof : (i) If $f \in D(H_o)$, we have

$$(f, H_o f) = \int |\underline{k}|^2 |\tilde{f}(\underline{k})|^2 d^\nu k \geq 0. \tag{2.30}$$

Hence $H_o \geq 0$. The unboundedness of H_o follows from Proposition 2.12.

(ii) Assume $H_o f = \lambda f$ for some $f \in D(H_o)$ and $\lambda \in \mathbb{R}$. Then

$$\int (\underline{k}^2 - \lambda)^2 |\tilde{f}(\underline{k})|^2 d^\nu k = 0,$$

so that $(|\underline{k}|^2 - \lambda)\tilde{f}(\underline{k}) = 0$ a.e. Since $\underline{k}^2 \neq \lambda$ except on a sphere in $\mathbb{R}^\nu$ (if $\lambda > 0$), which is a set of measure zero, we must have $\tilde{f}(\underline{k}) = 0$ a.e. Hence f is the zero vector, and thus H_o has no eigenvalue.

(iii) It is straightforward to check that multiplication by $(\underline{k}^2 - z)^{-1}$ is the inverse operator of $H_o - zI$. Since $|\underline{k}|^2 \in [0,\infty)$, the norm of this multiplication operator is equal to

$$\|(\underline{k}^2 - z)^{-1}\|_\infty = \sup_{\lambda \geq 0} |\lambda - z|^{-1},$$

which is finite if $z \notin [0,\infty)$. Thus $\sigma(H_o) \subseteq [0,\infty)$. On the other hand, if $z \in [0,\infty)$, multiplication by $(\underline{k}^2 - z)^{-1}$ is unbounded, though densely defined. Hence each $z \in [0,\infty)$ belongs to $\sigma(H_o)$.

(iv) $P_m(H_o - z)^{-1}$ is the multiplication operator by $k_m(\underline{k}^2 - z)^{-1}$ in $\tilde{L}^2(\mathbb{R}^\nu)$. By Proposition 2.12 :

$$\|P_m(H_o - z)^{-1}\| = \|k_m(\underline{k}^2 - z)^{-1}\|_\infty \leq \|(1 + |\underline{k}|^2)(\underline{k}^2 - z)^{-1}\|_\infty$$

$$= \sup_{\lambda \geq 0} \frac{1 + \lambda}{|\lambda - z|} < \infty \quad \text{if} \quad z \notin [0,\infty). \qquad \square$$

In the following proposition, we use the set $S(\mathbb{R}^\nu)$ of infinitely differentiable functions of rapid decrease, defined in the Appendix.

Proposition 2.18 : (a) If $f\in S(\mathbb{R}^\nu)$, then $f\in D(H_o)$ and

$$(H_o f)(\underline{x}) = -(\Delta f)(\underline{x}), \tag{2.31}$$

where $\Delta := \sum_{m=1}^{\nu} \partial^2/\partial x_m^2$ is the Laplacian.

(b) $(H_o + I)$ maps $S(\mathbb{R}^\nu)$ onto $S(\mathbb{R}^\nu)$.

(c) The restriction $\hat{H}_o$ of H_o to $S(\mathbb{R}^\nu)$ is essentially self-adjoint, and $\hat{H}_o^* = H_o$.

Proof : (a) By Lemma A.2, $S(\mathbb{R}^\nu)$ is invariant under Fourier transformation. Thus $f\in S(\mathbb{R}^\nu) \Rightarrow \tilde{f}\in S(\mathbb{R}^\nu)$, so that $|\tilde{f}(\underline{k})| \le c_n(1+|\underline{k}|)^{-n}$ for any $n > 0$. This implies that $f\in D(H_o)$, i.e. $S(\mathbb{R}^\nu)\subseteq D(H_o)$. Now for $f\in S(\mathbb{R}^\nu)$,

$$(H_o f)(\underline{x}) = (2\pi)^{-\nu/2} \int e^{i\underline{k}\cdot\underline{x}} |\underline{k}|^2 \tilde{f}(\underline{k}) d^\nu k =$$

$$= -(2\pi)^{-\nu/2} \int \tilde{f}(\underline{k})(\Delta e^{i\underline{k}\cdot\underline{x}}) d^\nu k = -\Delta[(2\pi)^{-\nu/2} \int \tilde{f}(\underline{k}) e^{i\underline{k}\cdot\underline{x}} d^\nu k]$$

$$= -(\Delta f)(\underline{x}),$$

since the derivatives may be interchanged with the integral.

(b) It is clear from the definition of $S(\mathbb{R}^\nu)$ in the Appendix that $f\in S(\mathbb{R}^\nu) \Rightarrow (\underline{k}^2 + 1)^\alpha \tilde{f}(\underline{k}) \in S(\mathbb{R}^\nu)$ for each $\alpha\in\mathbb{R}$. Taking $\alpha = \pm 1$, one obtains the result of (b).

(c) $\hat{H}_o$ is symmetric and positive, and $R(\hat{H}_o + I) = S(\mathbb{R}^\nu)$ is dense in $L^2(\mathbb{R}^\nu)$. Hence Proposition 2.4 implies the essential self-adjointness of $\hat{H}_o$, and also that its only self-adjoint extension is $\hat{H}_o^*$. Since H_o is a self-adjoint extension of $\hat{H}_o$, we must have $\hat{H}_o^* = H_o$. □

Remark : There are other linear submanifolds of $D(H_o)$ on which H_o is essentially self-adjoint. We mention here two such submanifolds : (a) The set $C_o^\infty(\mathbb{R}^\nu)$ of all infinitely differentiable functions of compact support (see [K,§V.5.2] for a proof), (b) the set $\tilde{C}_o^\infty(\mathbb{R}^\nu)$ of all functions $f:\mathbb{R}^\nu \to \mathbb{C}$ whose Fourier transform $\tilde{f}$ is infinitely differentiable and of compact support (the proof in this case is identical with the proof of Proposition 2.18 , replacing $S(\mathbb{R}^\nu)$ by $\tilde{C}_o^\infty(\mathbb{R}^\nu)$).

Problem 2.19 : Let $f \in S(\mathbb{R}^\nu)$. Then one has, for $m = 1,\ldots,\nu$:

$$(P_m f)(\underline{x}) = -i\frac{\partial f(\underline{x})}{\partial x_m} \quad \text{and} \quad (FQ_m f)(\underline{k}) = i\frac{\partial \tilde{f}(\underline{k})}{\partial k_m}.$$

Let φ and f be C^∞-functions form $\mathbb{R}^\nu$ to $\mathbb{C}$. Then, by Leibniz' rule for differentiating a product, we have

$$-\Delta[\varphi(\underline{x})f(\underline{x})] = -\varphi(\underline{x})(\Delta f)(\underline{x}) - (\Delta\varphi)(\underline{x})f(\underline{x})$$

$$- 2\sum_{m=1}^{\nu} \frac{\partial\varphi(\underline{x})}{\partial x_m}\frac{\partial f(\underline{x})}{\partial x_m}. \tag{2.32}$$

Let us denote by $\Phi \equiv \varphi(Q)$, $\Phi_{,m} \equiv \varphi_{,m}(Q)$ and $(\Delta\Phi) \equiv (\Delta\varphi)(Q)$ the multiplication operators in $L^2(\mathbb{R}^\nu)$ by $\varphi(\underline{x})$, $\varphi_{,m}(\underline{x}) := \partial\varphi(\underline{x})/\partial x_m$ and $(\Delta\varphi)(\underline{x})$ respectively. If we assume that φ, $\mathrm{grad}\,\varphi$ and $\Delta\varphi$ belong to $L^\infty(\mathbb{R}^\nu)$ and $f \in S(\mathbb{R}^\nu)$, then clearly the r.h.s. of (2.32) is in $L^2(\mathbb{R}^\nu)$. In view of Proposition 2.18 (a) and Problem 2.19, we then expect that (2.32) implies the following identity :

$$H_o\Phi f = \Phi H_o f - (\Delta\Phi)f - 2i\sum_{m=1}^{\nu} \Phi_{,m}P_m f. \tag{2.33}$$

By noticing that

$$\varphi_{,m}(\underline{x})\frac{\partial f(\underline{x})}{\partial x_m} = \frac{\partial}{\partial x_m}[\varphi_{,m}(\underline{x})f(\underline{x})] - \frac{\partial^2\varphi(\underline{x})}{\partial x_m^2}f(\underline{x}), \tag{2.34}$$

we similarly expect that (2.32) leads to

$$H_o \Phi f = \Phi H_o f + (\Delta\Phi) f - 2i \sum_{m=1}^{\nu} P_m \Phi_{,m} f. \tag{2.35}$$

It is the purpose of the next proposition to prove that (2.33) and (2.35) are in fact true for each $f \in D(H_o)$.

Proposition 2.20 : Let $\varphi : \mathbb{R}^\nu \to \mathbb{C}$ be a C^∞-function such that φ, grad φ and $\Delta\varphi$ are bounded, and let $f \in D(H_o)$. Then $\Phi f \in D(H_o)$, and $H_o \Phi f$ is given by (2.33) and (2.35).

Proof : (i) First assume that $f \in S(\mathbb{R}^\nu)$. Then $(\Phi f)(\underline{x})$, $(\Phi_{,m} f)(\underline{x})$ and $[(\Delta\Phi)f](\underline{x})$ are C^∞-functions that decrease rapidly as $|\underline{x}| \to \infty$. This allows us to integrate by parts in the integral below, where we assume $g \in S(\mathbb{R}^\nu)$ and use first Proposition 2.18 (a) :

$$(\hat{H}_o g, \Phi f) = -\int \overline{[\Delta g(\underline{x})]} (\Phi f)(\underline{x}) d^\nu x = -\int \overline{g(\underline{x})} \Delta (\Phi f)(\underline{x}) d^\nu x.$$

By (2.32), $-\Delta(\Phi f)(\cdot)$ defines a vector in $L^2(\mathbb{R}^\nu)$, which we denote by h. Therefore the preceding equation shows that $\Phi f \in D(\hat{H}_o^*) = D(H_o)$, and $\hat{H}_o^* \Phi f = H_o \Phi f = h$. Now h, defined as the r.h.s. of (2.32), is also given by the r.h.s. of (2.33), since $f \in S(\mathbb{R}^\nu)$. This proves (2.33) for $f \in S(\mathbb{R}^\nu)$.

If we assume that φ is such that $\varphi_{,m}(\cdot) f(\cdot) \in S(\mathbb{R}^\nu)$ for $m = 1,\ldots,\nu$ (e.g. that $\varphi(\underline{x}) = \text{const.}$ for $|\underline{x}| \geq R$), then (2.35) follows immediately by using (2.34). Without this assumption, one can show by an argument similar to the one given above that $\Phi_{,m} f \in D(P_m)$ and that $(iP_m \Phi_{,m} f)(\underline{x}) = \partial/\partial x_m [\varphi_{,m}(\underline{x}) f(\underline{x})]$, using the fact that P_m is essentially self-adjoint on $S(\mathbb{R}^\nu)$.

(ii) Now assume $f \in D(H_o)$, and let $g \in S(\mathbb{R}^\nu)$. Then $(H_o g, \Phi f) = (\Phi^* H_o g, f)$. Now $\Phi^* = \bar{\Phi}$, the multiplication operator by $\overline{\varphi(\underline{x})}$. We may use (2.35) to express $\bar{\Phi} H_o g$ and obtain

$$(H_o g,\Phi f) = (H_o\Phi^* g,f) - ((\Delta\Phi)^* g,f) - 2i\sum_{m=1}^{\nu}(P_m\Phi^*_{,m} g,f)$$

$$= (g,\Phi H_o f) - (g,(\Delta\Phi)f) - 2i\sum_{m=1}^{\nu}(g,\Phi_{,m}P_m f).$$

This shows that $\Phi f\in D(\hat{H}_o^*) = D(H_o)$ and that

$$\hat{H}_o^*\Phi f = \Phi H_o f - (\Delta\Phi)f - 2i\sum_{m=1}^{\nu}\Phi_{,m}P_m f.$$

Since $\hat{H}_o^* = H_o$, we have verified (2.33) for each $f\in D(H_o)$. The proof of (2.35) is similar. □

2.4 Perturbation Theory. Schrödinger Hamiltonians

We have seen that the Hamiltonian for a particle moving under the influence of a potential v is formally given as $H = H_o + V$. In the present section we give conditions on the potential v which allow one to consider the operator V as a "small" perturbation of H_o, so that the operator sum $H_o + V$ is self-adjoint. One sufficient condition clearly is the condition that $||v||_\infty < \infty$, since then $V\in\mathcal{B}(\mathcal{H})$, and the self-adjointness of $H_o + V$ follows from Problem 1.12 (d). However it is important to treat also unbounded potentials, since such potentials do occur in quantum mechanics (for instance the Coulomb potential $v(\underline{x}) = \gamma|\underline{x}|^{-1}$ describing the forces between charged particles).

We first give some abstract results, which will then be applied to Schrödinger operators. If A and B are self-adjoint and at least one of them, say B, is bounded, then A + B is self-adjoint with $D(A+B) = D(A)$. If both A and B are unbounded but $D(A+B) \equiv D(A)\cap D(B)$ is dense in $\mathcal{H}$, then A + B is symmetric but in general neither self-adjoint nor essentially self-adjoint. The symmetry of A + B is easy to verify : if $f,g\in D(A)\cap D(B)$, then

$(f,(A+B)g) = (f,Ag) + (f,Bg) = (Af,g) + (Bf,g) = ((A+B)f,g)$.

We now introduce the concept of relative boundedness which allows one to compare two unbounded operators.

Definition : Let A and B be linear operators. One says that B is A-bounded if

(i) $D(A) \subseteq D(B)$,

(ii) there are two numbers β and γ in $[0,\infty)$ such that

$$||Bf|| \leq \beta||Af|| + \gamma||f|| \qquad \forall\, f \in D(A). \tag{2.36}$$

The infimum of all numbers β for which (2.36) is true is called the A-bound of B.

Remark : The number γ in (2.36) may be different for different values of β. The A-bound of B is determined solely by considering all possible values of β.

Lemma 2.21 : Assume that $A = A^*$. (i) The following three statements are equivalent :

(a) B is A-bounded.

(b) $B(A-zI)^{-1} \in \mathcal{B}(\mathcal{H})$ for some $z \in \rho(A)$.

(c) $D(A) \subseteq D(B)$ and

$$||Bf||^2 \leq \beta_o^2||Af||^2 + \gamma_o^2||f||^2 \qquad \forall\, f \in D(A), \tag{2.37}$$

where β_o, γ_o are numbers in $[0,\infty)$.
The A-bound of B is also equal to the infimum of all numbers β_o for which (2.37) holds.

(ii) The following two statements are equivalent :

(d) B is A-bounded with A-bound $\upsilon < 1$.

(e) There exists a number $z\in\rho(A)$ such that $\|B(A-zI)^{-1}\| < 1$.

Proof : (i) (a) ⇒ (c) : Let $\eta > 0$. Then, for $f\in D(A)$,

$$(\eta^{1/2}\beta\|Af\| - \eta^{-1/2}\gamma\|f\|)^2$$

$$= \eta\beta^2\|Af\|^2 + \gamma^2\eta^{-1}\|f\|^2 - 2\beta\gamma\|Af\|\,\|f\| \geq 0. \tag{2.38}$$

Now, using first (2.36) and then (2.38), we get

$$\|Bf\|^2 \leq \beta^2\|Af\|^2 + \gamma^2\|f\|^2 + 2\beta\gamma\|Af\|\,\|f\|$$

$$\leq (1+\eta)\beta^2\|Af\|^2 + (1+\eta^{-1})\gamma^2\|f\|^2.$$

We therefore have (2.37) with $\beta_o = (1+\eta)^{1/2}\beta$, $\gamma_o = (1+\eta^{-1})^{1/2}\gamma$. Since η is an arbitrary positive number, the infimum of all possible β_o admitting an inequality (2.37) cannot be larger than the infimum of all β admitting an inequality (2.36) : Thus, if υ denotes the A-bound of B, we have $\upsilon \geq \inf \beta_o$.

(c) ⇒ (b) : Without loss of generality we may assume β_o, $\gamma_o > 0$. Then (2.13) allows us to rewrite (2.37) as

$$\|Bf\|^2 \leq \beta_o^2\|(A \pm i\gamma_o\beta_o^{-1})f\|^2. \tag{2.39}$$

Since $i\gamma_o\beta_o^{-1}\in\rho(A)$, $[A \pm i\gamma_o\beta_o^{-1}]^{-1}$ map $\mathcal{H}$ onto D(A), and we may set $f = (A - i\gamma_o\beta_o^{-1})^{-1}g$. Here, as f varies over D(A), g varies over $\mathcal{H}$. Inserting this expression for f into (2.39), one obtains

$$\|B(A - i\gamma_o\beta_o^{-1})^{-1}g\|^2 \leq \beta_o^2\|g\|^2 \qquad \forall\, g\in\mathcal{H}. \tag{2.40}$$

Hence $\|B(A - i\gamma_o\beta_o^{-1})^{-1}\| \leq \beta_o$, and this operator is defined everywhere since $i\gamma_o\beta_o^{-1}\in\rho(A)$ and $D(A)\subseteq D(B)$.

(b) ⇒ (a) : First notice that (b) implies $D(A)\subseteq D(B)$, since

$B(A-z)^{-1}$ is everywhere defined and $R((A-z)^{-1}) = D(A)$. Now let $f \in D(A)$. Then

$$||Bf|| = ||B(A-z)^{-1}(A-z)f||$$

$$\leq ||B(A-z)^{-1}|| \, ||Af|| + |z| \, ||B(A-z)^{-1}|| \, ||f||, \qquad (2.41)$$

which is (2.36) with $\beta = ||B(A-zI)^{-1}||$, $\gamma = |z|\beta$.

(2.40) and (2.41) imply that, given a number β_0 for which (2.37) holds, there is a number $\beta \leq \beta_0$ for which (2.36) is true. Hence $\upsilon \leq \inf \beta_0$. The opposite inequality has already been established, so that $\upsilon = \inf \beta_0$.

(ii) (e) ⇒ (d) : If $||B(A-zI)^{-1}|| < 1$, (2.41) gives

$$||Bf|| \leq \beta||Af|| + \gamma||f|| \quad \text{with} \quad \beta < 1.$$

(d) ⇒ (e) : If $\beta_0 < 1$, (2.40) implies that $||B(A-i\gamma_0\beta_0^{-1})^{-1}|| < 1$. □

Proposition 2.22 : If B is A-bounded, then $B(A-zI)^{-1} \in B(H)$ for each $z \in \rho(A)$.

Proof : Since $D(A) \subseteq D(B)$, $B(A-zI)^{-1}$ is everywhere defined in H. By the preceding lemma, there is a $z_0 \in \rho(A)$ such that $B(A-z_0)^{-1} \in B(H)$. Now by (2.15) and (2.16)

$$B(A-z)^{-1} = B(A-z_0)^{-1} + (z-z_0)B(A-z_0)^{-1}(A-z)^{-1},$$

which is in $B(H)$ if $z \in \rho(A)$. □

Proposition 2.23 (Kato-Rellich Theorem) : Let A be self-adjoint, B symmetric and A-bounded with A-bound $\upsilon < 1$. Then A + B is self-adjoint on D(A). Furthermore, if A is bounded from below, then so is A + B. (A is said to be bounded from below if $A + \mu \geq 0$, or equivalently if $(-\infty,-\mu)\in\rho(A)$, for some $\mu\in\mathbb{R}$.)

Proof : We shall not give the proof of the second part (boundedness from below). For the first part, we may assume that (2.37) holds with $0 < \beta_o < 1$ and $\gamma_o > 0$. We have already seen that $A + B$ is symmetric on $D(A) = D(A) \cap D(B)$.

Now let $f\in D(A)$. Then one has the identity

$$\left(\frac{\beta_o}{\gamma_o}(A+B) \pm i\right]f = [I + B(A \pm i\gamma_o\beta_o^{-1})^{-1}]\left(\frac{\beta_o}{\gamma_o}A \pm i\right)f.$$

Now $A = A^*$ implies that $\beta_o\gamma_o^{-1}A$ is self-adjoint. Therefore, by Proposition 2.2, $\beta_o\gamma_o^{-1}A \pm i$ map $D(A)$ onto $\mathcal{H}$. On the other hand, $\|B(A \pm i\gamma_o\beta_o^{-1})^{-1}\| \leq \beta_o < 1$ by (2.40). Hence $I + B(A \pm i\gamma_o\beta_o^{-1})^{-1}$ has an inverse in $\mathcal{B}(\mathcal{H})$, by Proposition 1.11. Hence $I + B(A \pm i\gamma_o\beta_o^{-1})^{-1}$ map $\mathcal{H}$ onto $\mathcal{H}$. This shows that

$$\mathcal{R}\left(\frac{\beta_o}{\gamma_o}(A+B) \pm i\right) = \mathcal{H},$$

so that, again by Proposition 2.2, $\beta_o\gamma_o^{-1}(A+B)$ is self-adjoint. It follows that $A + B$ is self-adjoint. □

Proposition 2.24 : Under the hypotheses of Proposition 2.23, B is $(A+B)$-bounded. Furthermore one has the second resolvent equation, for each $z\in\rho(A) \cap \rho(A+B)$:

$$(A+B-z)^{-1} = (A-z)^{-1} - (A-z)^{-1}B(A+B-z)^{-1}$$

$$= (A-z)^{-1} - (A+B-z)^{-1}B(A-z)^{-1}. \qquad (2.42)$$

Proof : (i) By (2.36), we have for all $f\in D(A) = D(A+B)$:

$$\|Bf\| \leq \beta\|Af\| + \gamma\|f\| = \beta\|(A+B)f - Bf\| + \gamma\|f\|$$

$$\leq \beta\|(A+B)f\| + \beta\|Bf\| + \gamma\|f\|.$$

Since $\beta < 1$, this implies that

$$||Bf|| \leq \beta(1-\beta)^{-1}||(A+B)f|| + \gamma(1-\beta)^{-1}||f||.$$

Thus B is (A + B)-bounded (but notice that its (A + B)-bound need not be less than 1 !).

(ii) Since A + B is defined on $D(A) = \mathcal{R}((A-z)^{-1})$, we have

$$B(A-z)^{-1} = [(A+B-z) - (A-z)](A-z)^{-1} = (A+B-z)(A-z)^{-1} - I.$$

Upon multiplication on the left by $(A+B-z)^{-1}$, one obtains the second identity in (2.42). The first one is derived similarly by changing the roles of A and A + B. (Notice that B is defined on $\mathcal{R}((A+B-z)^{-1}) = D(A)$.) □

Example 2.25 : Let $A = A^*$, $B = B^*$ and $B \in \mathcal{B}(\mathcal{H})$. Then we know from Problem 1.12 (d) that A + B is self-adjoint on D(A). This result of course also follows from Proposition 2.23. In fact, B is A-bounded with A-bound $\upsilon = 0$, since we may set $\beta = 0$, $\gamma = ||B||$ in (2.36).

In this example, (2.36) holds if β is set equal to the A-bound υ of B. In general this is not true. We shall see below that B may have A-bound $\upsilon = 0$ without being itself bounded (cf. Remark 2.29).

Example 2.26 : Let $A = A^*$ be unbounded, and let $B = -\lambda A$ with $\lambda \geq 0$. Then $||Bf|| \leq \lambda||Af||$ for each $f \in D(A)$, hence B is A-bounded with A-bound λ. If $\lambda < 1$, $A + B = (1-\lambda)A$, which is self-adjoint. On the other hand, if $\lambda = 1$, A + B is the restriction of the zero operator to D(A). This operator is not self-adjoint, because its adjoint is the zero operator with domain $\mathcal{H}$. This shows that the hypothesis $\upsilon < 1$ cannot be weakened in Proposition 2.23. (If $\upsilon = 1$, one can however show that A + B is essentially self-adjoint.)

We shall now apply Proposition 2.23 to Schrödinger operators in $L^2(\mathbb{R}^\nu)$. We begin with an auxiliary estimate.

Lemma 2.27 : Let $\mathcal{H} = L^2(\mathbb{R}^\nu)$. Let $2 \le p \le \infty$ and let $\varphi,\psi \in L^p(\mathbb{R}^\nu)$. Denote by $\varphi(\underline{P})$ the multiplication operator by $\varphi(\underline{k})$ in $\tilde{L}^2(\mathbb{R}^\nu)$ and by $\psi(\underline{Q})$ the multiplication operator by $\psi(\underline{x})$ in $L^2(\mathbb{R}^\nu)$, and define $A_{\varphi\psi} = \varphi(\underline{P})\psi(\underline{Q})$, $B_{\varphi\psi} = \psi(\underline{Q})\varphi(\underline{P})$. Then the closures of $A_{\varphi\psi}$ and $B_{\varphi\psi}$ are in $\mathcal{B}(L^2(\mathbb{R}^\nu))$, and

$$\|A_{\varphi\psi}\| \le \|\varphi\|_p \|\psi\|_p , \tag{2.43}$$

$$\|B_{\varphi\psi}\| \le \|\varphi\|_p \|\psi\|_p . \tag{2.44}$$

Remark : If $p = \infty$, the result is immediate from Proposition 2.12. If $p < \infty$, $A_{\varphi\psi}$ and $B_{\varphi\psi}$ are even compact operators. This will be shown in Lemma 3.13. For $p = 2$, $A_{\varphi\psi}$ and $B_{\varphi\psi}$ are Hilbert-Schmidt operators, and in this case the above lemma is a special case of Proposition 3.6.

Proof : The proof uses the Hölder inequality (A.4) and the fact that the Fourier transformation is a bounded operator from $L^q(\mathbb{R}^\nu)$ to $L^{q'}(\mathbb{R}^\nu)$ if $1 \le q \le 2$ and $q' = (1-q^{-1})^{-1}$:

$$\|\tilde{f}\|_{q'} \equiv \left(\int |\tilde{f}(\underline{k})|^{q'} d^\nu k\right)^{1/q'} \le \|f\|_q = \left(\int |f(\underline{x})|^q d^\nu x\right)^{1/q}, \tag{2.45}$$

see e.g. [SW, §V.1]. For $q = 2$, we know that (2.45) is an equality, since $\mathcal{F}$ is unitary in $L^2(\mathbb{R}^\nu)$. For $q = 1$, (2.45) is straightforward:

$$\|\tilde{f}\|_\infty = \sup_{k\in\mathbb{R}^\nu} |\tilde{f}(\underline{k})| \le (2\pi)^{-\nu/2} \int |f(\underline{x})| d^\nu x \le \|f\|_1 .$$

For general q one obtains (2.45) by interpolation.

(i) We show that $C_0^\infty(\mathbb{R}^\nu) \subseteq D(A_{\varphi\psi})$, so that $A_{\varphi\psi}$ is densely defined. Let $f \in C_0^\infty(\mathbb{R}^\nu)$ and let Δ be a bounded open set in $\mathbb{R}^\nu$ such that $f(\underline{x}) = 0 \ \forall\, \underline{x} \notin \Delta$. By the Hölder inequality, $\|\psi(\underline{Q})f\|_p \le \|\psi\|_p \|f\|_\infty < \infty$.

Hence, since $p \geq 2$: $\psi(\underline{Q})f \in L^r(\mathbb{R}^\nu)$ for each $r \in [1,2]$, by Lemma A.1 (a). If we now use (2.45), we find that $F\psi(\underline{Q})f \in L^s(\mathbb{R}^\nu)$ for each $s \in [2,\infty]$, in particular for $s_0 = (1/2 - p^{-1})^{-1}$. Thus, since $1/2 = p^{-1} + s_0^{-1}$, $\varphi(\underline{k})[F\psi(\underline{Q})f](\underline{k}) \in \tilde{L}^2(\mathbb{R}^\nu)$, in other words $\varphi(\underline{P})\psi(\underline{Q})f \in L^2(\mathbb{R}^\nu)$.

(ii) We now estimate the norm of $A_{\varphi\psi}$. Let $g \in L^2(\mathbb{R}^\nu)$ and $q = (1/2 + p^{-1})^{-1}$. Notice that $1 \leq q \leq 2$. By (2.45) and the Hölder inequality :

$$\| F\psi(\underline{Q})g\|_{q'} \leq \|\psi(\underline{Q})g\|_q \leq \|\psi\|_p \|g\|_2.$$

Now notice that $p^{-1} + (q')^{-1} = p^{-1} + 1 - q^{-1} = 1/2$. Hence $\varphi(\underline{k})[F\psi(\underline{Q})g](\underline{k}) \in \tilde{L}^2(\mathbb{R}^\nu)$ by the Hölder inequality, and

$$\|A_{\varphi\psi}g\| = \|\varphi(\underline{k})[F\psi(\underline{Q})g](\underline{k})\|_2 \leq \|\varphi\|_p \|F\psi(\underline{Q})g\|_{q'}$$

$$\leq \|\varphi\|_p \|\psi\|_p \|g\|_2 ,$$

where we have used the unitarity of F in $L^2(\mathbb{R}^\nu)$ to obtain the first equality. This proves (2.43). The proof of (2.44) is similar by inverting the roles of $L^2(\mathbb{R}^\nu)$ and $\tilde{L}^2(\mathbb{R}^\nu)$. □

We now introduce a class of potentials that will often be used in these notes. A measurable function $v : \mathbb{R}^\nu \to \mathbb{R}$ will be said to be of class $V_\emptyset$ of it can be written as $v = v_1 + v_2$ with $v_1 \in L^\infty(\mathbb{R}^\nu)$ and $v_2 \in L^p(\mathbb{R}^\nu)$ for some p satisfying $p \geq 2$ and $p > \nu/2$.

Examples : Some examples from quantum mechanics, for $\nu = 3$, are the following potentials :

i) Square well or barrier :

$$v(\underline{x}) = \begin{cases} v_0 & |\underline{x}| \leq a \\ 0 & |\underline{x}| > a, \end{cases} \tag{2.46}$$

where $v_0 \in \mathbb{R}$. We may take $v_1 = v$, $v_2 = 0$ or $v_1 = 0$, $v_2 = v$.

ii) Yukawa potential :

$$v(\underline{x}) = \alpha|\underline{x}|^{-1} \exp(-\mu|\underline{x}|) \tag{2.47}$$

with $\alpha \in \mathbb{R}$ and $\mu > 0$. Here one may take $v_1 = 0$, $v_2 = v$ and $p = 2$.

iii) Coulomb potential :

$$v(\underline{x}) = \alpha|\underline{x}|^{-1}, \qquad \alpha \in \mathbb{R}. \tag{2.48}$$

Here one takes $v_1(\underline{x}) = v(\underline{x})$ if $|\underline{x}| > 1$ and $v_1(\underline{x}) = 0$ if $|\underline{x}| \le 1$, $v_2(\underline{x}) = v(\underline{x}) - v_1(\underline{x})$ and $p = 2$.

Proposition 2.28 : Let $\mathcal{H} = L^2(\mathbb{R}^\nu)$, $\nu = 1,2,\ldots$. Let $v \in \mathcal{V}_\emptyset$. Then $D(H_o) \subseteq D(V)$, V is H_o-bounded with H_o-bound $\upsilon = 0$, and $V(H_o - z)^{-1} \in \mathcal{B}(\mathcal{H})$ for each $z \in \rho(H_o)$.

Proof : By Lemma 2.21, in particular Eq. (2.41), and Proposition 2.22, it suffices to show that

$$\lim_{\mu\to\infty} \|V(H_o + \mu)^{-1}\| = 0 \tag{2.49}$$

(notice that $(-\infty,0) \in \rho(H_o)$). Now

$$\begin{aligned} \|V(H_o + \mu)^{-1}\| &\le \|V_1\|\,\|(H_o + \mu)^{-1}\| + \|V_2(\underline{P}^2 + \mu)^{-1}\| \\ &\le \|v_1\|_\infty \sup_{\lambda\ge 0}|\lambda + \mu|^{-1} + \|v_2\|_p \|(\underline{k}^2 + \mu)^{-1}\|_p , \end{aligned} \tag{2.50}$$

where we have used Proposition 2.12 for the first term and Lemma 2.27 for the second term. Notice that $(\underline{k}^2 + \mu)^{-1}$ is in $L^p(\mathbb{R}^\nu)$ under our assumptions on p, and it is majorized, for all $\mu \ge 1$, by $(\underline{k}^2 + 1)^{-1}$. Hence, by the Lebesgue dominated convergence theorem, $\lim \|(\underline{k}^2 + \mu)^{-1}\|_p = 0$ as $\mu \to \infty$. Since $|\lambda + \mu|^{-1} \le \mu^{-1}$ if $\lambda, \mu \ge 0$, we see that (2.50) implies (2.49). □

Remark 2.29 : (2.49) shows that the H_o-bound of V is $\upsilon = 0$. However, if v is unbounded, it is not possible to set $\beta = 0$ in (2.36), although β may be taken arbitrarily small. It is clear that, the smaller one chooses β, the larger one has to take γ.

Proposition 2.30 : Under the assumptions of Proposition 2.28, $H = H_o + V$ is self-adjoint and bounded from below.

Proof : This follows immediately from Propositions 2.23, 2.28 and 2.17 (a). □

Lemma 2.31 : Let $\mathcal{H} = L^2(\mathbb{R}^\nu)$. Assume that $v,w \in \mathcal{V}_\emptyset$, let $H = H_o + V$ and denote by W the multiplication operator by $w(\underline{x})$. Then, for each $z \in \rho(H)$, the operator $W(H-z)^{-1}$ and the closure of $(H-z)^{-1}W$ belong to $\mathcal{B}(\mathcal{H})$.

Proof : (i) Let $z \in \rho(H)$. Then

$$W(H-z)^{-1} = W(H_o+1)^{-1}(H_o+1)(H-z)^{-1}$$
$$= W(H_o+1)^{-1}[I - V(H-z)^{-1} + (1+z)(H-z)^{-1}].$$

Now $W(H_o+1)^{-1} \in \mathcal{B}(\mathcal{H})$ by Proposition 2.28, and $V(H-z)^{-1} \in \mathcal{B}(\mathcal{H})$ by Propositions 2.28, 2.24 and 2.22. Hence $W(H-z)^{-1} \in \mathcal{B}(\mathcal{H})$.

(ii) We have $D(H_o) = D(H) \subseteq D(W)$ by Proposition 2.28. This shows that $(H-z)^{-1}W$ is densely defined. Since $(H-z)^{-1*} = (H-\bar{z})^{-1}$ (see (4.48)), we have for $g \in D(W)$ and $f \in \mathcal{H}$:

$$(f,(H-z)^{-1}Wg) = (W(H-\bar{z})^{-1}f,g) = (f,[W(H-\bar{z})^{-1}]^*f).$$

Hence $(H-z)^{-1}Wg = [W(H-\bar{z})^{-1}]^*g$. By (i) and (1.34), $||(H-z)^{-1}Wg|| \leq ||W(H-\bar{z})^{-1}||\,||g||$. Thus $(H-z)^{-1}W$ is bounded and densely defined, i.e. its closure is in $\mathcal{B}(\mathcal{H})$ (see Proposition 1.7). □

2.5 Schrödinger Operators with Singular Potentials

We shall later consider Hamiltonians with very singular potentials. These potentials may be singular at infinity or have much stronger local singularities than those allowed by Proposition 2.28 (for example $v(\underline{x}) = \alpha|\underline{x}|^{-k}$, where k may be any positive number). In the present section we study domain properties of such Hamiltonians.

We assume that the local singularities of the potential v are restricted to a set Γ which is a closed set of measure zero, and that the (real-valued) potential satisfies

$$v \in L^p_{loc}(\mathbb{R}^\nu\backslash\Gamma), \text{ with } p \geq 2,\ p > \nu/2. \tag{2.51}$$

This means that $\mathbb{R}^\nu\backslash\Gamma$ is an open set and that, for each compact subset Δ of $\mathbb{R}^\nu\backslash\Gamma$,

$$\int_\Delta |v(\underline{x})|^p d^\nu x < \infty, \tag{2.52}$$

or equivalently that

$$\int |\varphi(\underline{x})v(\underline{x})|^p d^\nu x < \infty \quad \text{for each} \quad \varphi \in C_0^\infty(\mathbb{R}^\nu\backslash\Gamma), \tag{2.53}$$

where $C_0^\infty(\mathbb{R}^\nu\backslash\Gamma)$ denotes the set of all infinitely differentiable functions each of which vanishes in some neighbourhood of Γ and of infinity. (In the example given above, Γ consists of a single point $\underline{x} = 0$. Another example is a potential which is singular on a sphere, e.g. $v(\underline{x}) = (|\underline{x}| - 1)^{-k}$ $(k > 0)$, or at infinity, e.g. $v(\underline{x}) = |\underline{x}|^n$ $(n > 0)$.)

To define H, we first introduce the so-called minimal operator $\hat{H}$ as follows :

$$D(\hat{H}) = C_0^\infty(\mathbb{R}^\nu\backslash\Gamma), \quad (\hat{H}f)(\underline{x}) = -\Delta f(\underline{x}) + v(\underline{x})f(\underline{x}). \tag{2.54}$$

Clearly $-\Delta f(\underline{x}) = (H_0 f)(\underline{x})$ is in $L^2(\mathbb{R}^\nu)$. To see that $Vf \in L^2(\mathbb{R}^\nu)$, choose a function $\varphi \in C_0^\infty(\mathbb{R}^\nu \setminus \Gamma)$ such that $\varphi(\underline{x})f(\underline{x}) = f(\underline{x})$ for all $\underline{x}$ (cf. Lemma A.4 (b)). Then $Vf = V\varphi(\underline{Q})f = V\varphi(\underline{Q})(H_0 + 1)^{-1}(H_0 + 1)f$, which is in $L^2(\mathbb{R}^\nu)$, since $f \in D(H_0)$ and $V\varphi(\underline{Q})(H_0 + 1)^{-1} \in \mathcal{B}(L^2(\mathbb{R}^\nu))$ by the fact that $V\varphi(\underline{Q})$ is H_0-bounded.

Since v is real-valued and $C_0^\infty(\mathbb{R}^\nu \setminus \Gamma)$ is dense in $L^2(\mathbb{R}^\nu)$ (cf. Lemma A.3), $\hat{H}$ is a symmetric operator. It can be shown that it always has self-adjoint extensions. In general $\hat{H}$ is not essentially self-adjoint and may have an uncountable number of self-adjoint extensions (see Problem 2.39 for an example). In what follows we let H be an arbitrary self-adjoint extension of $\hat{H}$, and we shall give a characterization of vectors in the domain of H valid for each self-adjoint extension H.

Intuitively it is clear what to expect. We saw in Proposition 2.30 that, if $v \in L^p(\mathbb{R}^\nu)$ with $p \geq 2$, $p > \nu/2$, then $H = H_0 + V$ is self-adjoint with $D(H) = D(H_0)$. In analogy with this one expects that, if $v \in L^p_{loc}(\mathbb{R}^\nu \setminus \Gamma)$ with p as before, the domains of a self-adjoint extension H and of H_0 should be the same locally on $\mathbb{R}^\nu \setminus \Gamma$. In mathematical terms, this would be expressed as follows : If $f \in D(H)$, then $\varphi(\underline{Q})f \in D(H_0)$ for each $\varphi \in C_0^\infty(\mathbb{R}^\nu \setminus \Gamma)$, and if $g \in D(H_0)$, then $\varphi(\underline{Q})g \in D(H)$ for each $\varphi \in C_0^\infty(\mathbb{R}^\nu \setminus \Gamma)$. Furthermore, if v is in L^p near infinity, it should not be necessary to require that $\varphi(\underline{x}) = 0$ near infinity. We shall now proceed to prove these results.

<u>Proposition 2.32</u> : Let v satisfy (2.51), let H be a self-adjoint extension of $\hat{H}$, and let $\varphi \in C^\infty(\mathbb{R}^\nu)$ be such that φ, gradφ and $\Delta\varphi$ are in $L^\infty(\mathbb{R}^\nu)$ and $v\varphi = w_1 + w_2$ with $w_1 \in L^\infty(\mathbb{R}^\nu)$, $w_2 \in L^q(\mathbb{R}^\nu)$ with $q \geq 2$, $q > \nu/2$ (i.e. $v\varphi \in \mathcal{V}_\emptyset$). Then, for each $z \in \rho(H_0)$, $H\varphi(\underline{Q})(H_0 - z)^{-1} \in \mathcal{B}(\mathcal{H})$. Furthermore $H\varphi(\underline{Q})(H_0 - z)^{-1} = H_0\varphi(\underline{Q})(H_0 - z)^{-1} + V\varphi(\underline{Q})(H_0 - z)^{-1}$.

Remark : If v is singular on Γ, φ must be zero in a neighbourhood of Γ. Similarly, if v is singular at infinity, φ must be zero near infinity (at least in a neighbourhood of those directions along which v has singularities). On the other hand, if $v = v_1 + v_2$ near infinity with $v_1 \in L^\infty$, $v_2 \in L^q$, then φ need not tend to zero at infinity but must remain bounded there.

Proof : We use the notation introduced before Proposition 2.20. Thus for instance Φ denotes the multiplication operator by $\varphi(\underline{x})$.

(i) First assume $\varphi \in C_o^\infty(\mathbb{R}^\nu \backslash \Gamma)$. Then Φ maps $S(\mathbb{R}^\nu)$ into $D(\hat{H})$. Since $\hat{H} \subseteq H$, we have $H \subseteq \hat{H}^*$. Hence, for each $f \in S(\mathbb{R}^\nu)$ and $g \in D(H)$:

$$(Hg,\Phi f) = (\hat{H}^* g,\Phi f) = (g,\hat{H}\Phi f) = (g,Tf),$$

where T is given, from Proposition 2.20, by

$$T = \Phi H_o + V\Phi - (\Delta\Phi) - 2i \sum_{m=1}^{\nu} \Phi_{,m} P_m . \tag{2.55}$$

Since f is in the domain of each of the operators on the r.h.s., we have $\Phi f \in D(H)$ and $H\Phi f = Tf$.

Now by Proposition 2.18 (b), we may write $f = (H_o + I)^{-1}h$, and, if f varies over $S(\mathbb{R}^\nu)$, then h also varies over $S(\mathbb{R}^\nu)$. Therefore $H\Phi(H_o + I)^{-1}$ is defined on the dense set $S(\mathbb{R}^\nu)$. This operator is also bounded, since by (2.55)

$$H\Phi(H_o + I)^{-1}h = \Phi h - \Phi(H_o + I)^{-1}h + V\Phi(H_o + I)^{-1}h$$

$$- (\Delta\Phi)(H_o + I)^{-1}h - 2i \sum_{m=1}^{\nu} \Phi_{,m} P_m (H_o + I)^{-1}h, \tag{2.56}$$

and each of the operators occuring on the r.h.s. is in $B(L^2(\mathbb{R}^\nu))$ (cf. Propositions 2.28 and 2.17 (b)).

(ii) If φ is not of compact support, choose a function $\chi \in C_o^\infty(\mathbb{R}^\nu)$

such that

$$0 \leq \chi(\underline{x}) \leq 1 \ \forall \underline{x} \in \mathbb{R}^\nu, \ \chi(\underline{x}) = 1 \text{ for } |\underline{x}| \leq 1, \ \chi(\underline{x}) = 0 \text{ for } |\underline{x}| \geq 2. \tag{2.57}$$

Set $\chi_n(\underline{x}) = \chi(\underline{x}/n)$, $\varphi_n(\underline{x}) = \chi_n(\underline{x})\varphi(\underline{x})$ and let Φ_n be the multiplication operator by $\varphi_n(\underline{x})$. Notice that

$$\frac{\partial \chi_n(\underline{x})}{\partial x_m} = \frac{1}{n}\left(\frac{\partial \chi}{\partial x_m}\right)\left(\frac{\underline{x}}{n}\right), \qquad (\Delta\chi_n)(\underline{x}) = \frac{1}{n^2}(\Delta\chi)\left(\frac{\underline{x}}{n}\right). \tag{2.58}$$

We have

$$||\Phi_n f - \Phi f||^2 \leq \int_{|\underline{x}|\geq n} |\varphi(\underline{x})|^2 |f(\underline{x})|^2 d^\nu x \leq ||\varphi||_\infty^2 \int_{|\underline{x}|\geq n} |f(\underline{x})|^2 d^\nu x,$$

which tends to zero as $n \to \infty$. Hence s-lim $\Phi_n = \Phi$ as $n \to \infty$. Similarly, by (2.58)

$$||\left(\frac{\partial \varphi_n}{\partial x_m}\right)(\underline{Q}) f - \left(\frac{\partial \varphi}{\partial x_m}\right)(\underline{Q}) f||$$

$$\leq ||\frac{\partial \varphi}{\partial x_m}||_\infty \left(\int_{|\underline{x}|\geq n} |f(\underline{x})|^2 d^\nu x\right)^{1/2} + \frac{1}{n}||\frac{\partial \chi}{\partial x_m}||_\infty ||\varphi||_\infty ||f||,$$

which also tends to zero as $n \to \infty$, implying that $(\partial\varphi_n/\partial x_m)(\underline{Q}) \to$ $\to (\partial\varphi/\partial x_m)(\underline{Q})$ strongly as $n \to \infty$. A similar estimate gives $(\Delta\varphi_n)(\underline{Q}) \to (\Delta\varphi)(\underline{Q})$ strongly.

Now $\varphi_n \in C_o^\infty(\mathbb{R}^\nu \backslash \Gamma)$, so that the identity (2.56) holds true if Φ is replaced by Φ_n. As $n \to \infty$, the r.h.s. of (2.56) converges to the same expression for the operator Φ, since by Lemma 2.27

$$||V\Phi(H_o + I)^{-1}h - V\Phi_n(H_o + I)^{-1}h|| \leq ||w_1||_\infty ||[I - \chi_n(\underline{Q})](H_o + I)^{-1}h||$$

$$+ \left(\int_{|\underline{x}|\geq n} |w_2(\underline{x})|^q d^\nu x\right)^{1/q} ||(\underline{k}^2 + 1)^{-1}||_q ||h||,$$

which converges to zero as $n \to \infty$.

Thus we have the following : For $h\in S(\mathbb{R}^\nu)$, $\Phi_n(H_0+I)^{-1}h\in D(H)$, $\Phi_n(H_0+I)^{-1}h \to \Phi(H_0+I)^{-1}h$ strongly and $H\Phi_n(H_0+I)^{-1}h$ converges strongly to the r.h.s. of (2.56) as $n\to\infty$. Since $H = H^*$, H is closed; hence $\Phi(H_0+I)^{-1}h\in D(H)$ and $H\Phi(H_0+I)^{-1}h$ is given by the r.h.s. of (2.56). Thus (2.56) holds also in this case, for each $h\in S(\mathbb{R}^\nu)$.

(iii) Next we show, using (2.56), that $H\Phi(H_0+I)^{-1}$ is defined everywhere. For fixed $g\in H$, set $f = \Phi(H_0+I)^{-1}g$ and choose a sequence $\{g_n\}\in S(\mathbb{R}^\nu)$ such that s-lim $g_n = g$. If we set $f_n = \Phi(H_0+I)^{-1}g_n$, then s-lim $f_n = f$. By (2.56), $\{Hf_n\}$ is strongly Cauchy :

$$||Hf_n - Hf_m|| \leq ||T(H_0+I)^{-1}||\,||f_n - f_m|| \to 0 \quad \text{as} \quad m,n\to\infty.$$

Using again the closedness of H, we see that $f\in D(H)$ and that $Hf = H\Phi(H_0+I)^{-1}g = T(H_0+I)^{-1}g$. Thus $H\Phi(H_0+I)^{-1} = T(H_0+I)^{-1}$, which is in $B(H)$ as pointed out in (2.56).

By combining the preceding result with the first resolvent equation (2.15), one sees that $H\Phi(H_0-z)^{-1}\in B(H)$ for each $z\in\rho(H_0)$.

(iv) Let $g\in H$ and set $h = H\Phi(H_0-z)^{-1}g - H_0\Phi(H_0-z)^{-1}g - V\Phi(H_0-z)^{-1}g$ (notice that $\Phi(H_0-z)^{-1}g\in D(H_0)\cap D(V)$ by Propositions 2.20 and 2.28, since $v\varphi\in V_\emptyset$). We must show that $h = \theta$. For this, let $f\in D(\hat{H})$. Then $f\in D(H)\cap D(H_0)\cap D(V)$ and $Hf = H_0f + Vf$. Hence $(f,h) = (Hf - H_0f - Vf, \Phi(H_0-z)^{-1}g) = 0$ $\forall f\in D(\hat{H})$. Since $D(\hat{H})$ is dense in $L^2(\mathbb{R}^\nu)$, we have $h = \theta$ by Lemma 1.5. □

Lemma 2.33 : Let $w\in L^p(\mathbb{R}^\nu)$, with $p\geq 2$, $p > \nu/2$. Let $a,b\geq 0$ be such that $a + b > \nu/(2p)$. Then the closure of $(H_0+I)^{-a}w(\underset{\sim}{Q})(H_0+I)^{-b}$ is in $B(L^2(\mathbb{R}^\nu))$.

Proof : (i) The operator $(H_0+I)^{-a}w(\underset{\sim}{Q})(H_0+I)^{-b}$ is densely defined, e.g. on $S(\mathbb{R}^\nu)$. Indeed

$$w(\underset{\sim}{Q})(H_o + I)^{-b}f = w(\underset{\sim}{Q})(H_o + I)^{-1}(H_o + I)^{1-b}f.$$

Here, if $f\in S(\mathbb{R}^\nu)$, then $f\in D((H_o + I)^{1-b})$, whereas $w(\underset{\sim}{Q})(H_o + I)^{-1}\in B(L^2(\mathbb{R}^\nu))$ by Proposition 2.28.

(ii) Let $w_1(\underset{\sim}{x}) = |w(\underset{\sim}{x})|^{a/(a+b)}$ and $w_2(\underset{\sim}{x}) = |w(\underset{\sim}{x})|^{b/(a+b)}\,\mathrm{sign}\,w(\underset{\sim}{x})$. Then $w = w_1w_2$, and $w_1\in L^{p(a+b)/a}(\mathbb{R}^\nu)$, $w_2\in L^{p(a+b)/b}(\mathbb{R}^\nu)$. Furthermore $\underset{\sim}{k} \mapsto (\underset{\sim}{k}^2 + 1)^{-a}\in L^{p(a+b)/a}(\mathbb{R}^\nu)$ and $\underset{\sim}{k} \mapsto (\underset{\sim}{k}^2 + 1)^{-b}\in L^{p(a+b)/b}(\mathbb{R}^\nu)$. Hence, by Lemma 2.27, the closures of $(H_o + I)^{-a}w_1(\underset{\sim}{Q})$ and $w_2(\underset{\sim}{Q})(H_o + I)^{-b}$ are in $B(L^2(\mathbb{R}^\nu))$. Hence the densely defined operator

$$(H_o + I)^{-a}w(\underset{\sim}{Q})(H_o + I)^{-b} = [(H_o + I)^{-a}w_1(\underset{\sim}{Q})][w_2(\underset{\sim}{Q})(H_o + I)^{-b}]$$

has a bounded extension, and the result of the lemma follows from Proposition 1.7. □

<u>Proposition 2.34</u> : Let $H = L^2(\mathbb{R}^\nu)$, $v\in L^p_{loc}(\mathbb{R}^\nu\backslash\Gamma)$ with $p \geq 2$, $p > \nu/2$, and let H be a self-adjoint extension of $\hat{H}$. Then $H_o\varphi(\underset{\sim}{Q})(H - z)^{-1}\in B(L^2(\mathbb{R}^\nu))$ for each $\varphi\in C_o^\infty(\mathbb{R}^\nu\backslash\Gamma)$ and each $z\in\rho(H)$.

<u>Proof</u> : We again set $\Phi = \varphi(\underset{\sim}{Q})$. Notice that $\Phi^* = \bar{\varphi}(\underset{\sim}{Q})$, where $\bar{\varphi}(\underset{\sim}{x}) := \overline{\varphi(x)}$.

Let $h\in S(\mathbb{R}^\nu)$, $g\in H$ and set $f = (H - i)^{-1}g$. Since $\Phi^*h\in C_o^\infty(\mathbb{R}^\nu\backslash\Gamma) \subseteq D(\hat{H})$ and $f\in D(H) \subseteq D(\hat{H}^*)$, we obtain by using (2.33) that

$$(\Phi f,(H_o + I)h) = (f,\Phi^*(H_o + I)h)$$

$$= (f,H_o\Phi^*h) + (f,(\Delta\Phi)^*h) + 2i\sum_{m=1}^{\nu}(f,\Phi^*_{,m}P_mh) + (f,\Phi^*h)$$

$$= (\Phi\hat{H}^*f,h) - (f,V\Phi^*h) + ((\Delta\Phi)f,h)$$

$$+ 2i\sum_{m=1}^{\nu}(\Phi_{,m}f,P_mh) + (\Phi f,h). \qquad (2.59)$$

Let $\psi \in C_0^\infty(\mathbb{R}^\nu \setminus \Gamma)$ be such that $\psi(\underline{x}) = 1$ for all $\underline{x}$ for which $\varphi(\underline{x}) \neq 0$, i.e. such that $\psi(\underline{x})\varphi(\underline{x}) = \varphi(\underline{x})$ (cf. Lemma A.4 (b)). Then $(f, V\Phi^* h) = (\psi(\underline{Q})f, V\Phi^* h)$. Also, $\hat{H}^* f = Hf$, since $\hat{H}^*$ is an extension of H. We insert this into (2.59) and (proceeding formally for a moment) rewrite (2.59) as follows :

$$(\Phi f, (H_0 + I)h) = (e, (H_0 + I)^a h) \tag{2.60}$$

with

$$e = (H_0 + I)^{-a}\Phi H f - [(H_o + I)^{-a} V\Phi(H_0 + I)^{-b}](H_0 + I)^b \psi(\underline{Q}) f$$

$$+ (H_0 + I)^{-a}(\Delta\Phi) f - 2i \sum_{m=1}^{\nu} P_m (H_0 + I)^{-(a+b)} (H_0 + I)^b \Phi_{,m} f$$

$$+ (H_0 + I)^{-a} \Phi f. \tag{2.61}$$

If the r.h.s. of (2.61) defines a vector in $L^2(\mathbb{R}^\nu)$ and $0 \leq a \leq 1$, we have from (2.60) that

$$(\Phi f, (H_0 + I)h) = ((H_0 + I)^{-(1-a)} e, (H_0 + I)h). \tag{2.62}$$

Since $(H_0 + I)\mathcal{S}(\mathbb{R}^\nu)$ is dense in $\mathcal{H}$, this implies by Lemma 1.5 that $\Phi f = (H_0 + I)^{-(1-a)} e$, in other words that $\Phi f \in D((H_0 + I)^{1-a})$ and

$$(H_0 + I)^{1-a} \Phi (H - i)^{-1} g = (H_0 + I)^{1-a} \Phi f = e. \tag{2.63}$$

Furthermore, by (2.61), we then have

$$\|(H_0 + I)^{1-a} \Phi (H - i)^{-1}\| \leq \|(H_0 + I)^{-a}\| \, \|\Phi\| \, \|H(H - i)^{-1}\|$$

$$+ \|(H_0 + I)^{-a} V\Phi(H_0 + I)^{-b}\| \, \|(H_0 + I)^b \psi(\underline{Q})(H - i)^{-1}\|$$

$$+ \|(H_0 + I)^{-a}\| \, \|(\Delta\Phi)\| \, \|(H - i)^{-1}\| + \|(H_0 + I)^{-a}\| \, \|\Phi\| \, \|(H - i)^{-1}\|$$

$$+ 2 \sum_{m=1}^{\nu} \|P_m (H_0 + I)^{-(a+b)}\| \, \|(H_0 + I)^b \Phi_{,m} (H - i)^{-1}\|. \tag{2.64}$$

Clearly $\|\Phi\| < \infty$ and $\|(\Delta\Phi)\| < \infty$. Also $\|(H-i)^{-1}\| \leq 1$ and $\|H(H-i)^{-1}\| = \|I+i(H-i)^{-1}\| \leq 2$. We choose a and b such that $a\in[0,1]$, $b \geq 0$, $a + b > 1/2$ and $a + b > \nu/(2p)$. Then $\|(H_o+I)^{-a}\| \leq 1$, $\|P_m(H_o+I)^{-(a+b)}\| < \infty$ by Proposition 2.12 and $\|(H_o+I)^{-a}V\Phi(H_o+1)^{-b}\| < \infty$ by Lemma 2.33.

Since $\psi, \partial\varphi/\partial x_m \in C_o^\infty(\mathbb{R}^\nu\setminus\Gamma)$, (2.64) now implies the following : Let $a\in[0,1]$, $b \geq 0$, $a + b > \max(1/2,\nu/2p)$ and suppose we know that $(H_o+I)^b\chi(\underline{Q})(H-i)^{-1}\in\mathcal{B}(\mathcal{H})$ for all $\chi\in C_o^\infty(\mathbb{R}^\nu\setminus\Gamma)$. Then $(H_o+I)^{1-a}\varphi(\underline{Q})(H-i)^{-1}\in\mathcal{B}(\mathcal{H})$ for each $\varphi\in C_o^\infty(\mathbb{R}^\nu\setminus\Gamma)$.

We use this iteratively to show that $(H_o+I)\varphi(\underline{Q})(H-i)^{-1}\in\mathcal{B}(\mathcal{H})$. Since $\nu/(2p) < 1$, there exists an integer m such that $1 - 1/m > \max(1/2,\nu/2p)$. First choose $b = 0$, $a = 1 - 1/m$ to obtain that $(H_o+I)^{1/m}\varphi(\underline{Q})(H-i)^{-1}\in\mathcal{B}(\mathcal{H})$ for each $\varphi\in C_o^\infty(\mathbb{R}^\nu\setminus\Gamma)$. Next choose $b = 1/m$, $a = 1 - 2/m$ to find that $(H_o+I)^{2/m}\varphi(\underline{Q})(H-i)^{-1}\in\mathcal{B}(\mathcal{H})$. After $m - 2$ more such steps, one arrives at $(H_o+I)\varphi(\underline{Q})(H-i)^{-1}\in$ $\in\mathcal{B}(\mathcal{H})$, which proves the result of the proposition for $z = i$. By combining this with the first resolvent equation (2.15), one obtains $(H_o+I)\varphi(\underline{Q})(H-z)^{-1}\in\mathcal{B}(\mathcal{H})$ for each $z\in\rho(H)$. □

<u>Corollary 2.35</u> : Assume the hypotheses of Proposition 2.34. Let $f\in D(\hat{H}^*)$ and $\varphi\in C_o^\infty(\mathbb{R}^\nu\setminus\Gamma)$. Then $\varphi(\underline{Q})f\in D(H_o)\cap D(H)\cap D(V)$ and $H\varphi(\underline{Q})f = H_o\varphi(\underline{Q})f + V\varphi(\underline{Q})f$.

<u>Proof</u> : Assume $f\in D(\hat{H}^*)$ instead of $f\in D(H)$ in the preceding proof. One then gets : If $a\in[0,1]$, $b \geq 0$, $(a+b) > \max(1/2,\nu/2p)$, then $\chi(\underline{Q})f\in D((H_o+I)^b)$ for each $\chi\in C_o^\infty(\mathbb{R}^\nu\setminus\Gamma)$ implies $\varphi(\underline{Q})f\in D((H_o+I)^{1-a})$ for each $\varphi\in C_o^\infty(\mathbb{R}^\nu\setminus\Gamma)$. Beginning with $b = 0$ and iterating as above, one readily obtains that $\varphi(\underline{Q})f\in D(H_o)$.

Next, let $\psi\in C_o^\infty(\mathbb{R}^\nu\setminus\Gamma)$ be such that $\psi\varphi = \varphi$, cf. Lemma A.4 (b). Then

$$\varphi(\underset{\sim}{Q})f = \psi(\underset{\sim}{Q})\varphi(\underset{\sim}{Q})f = \psi(\underset{\sim}{Q})(H_o + I)^{-1}(H_o + I)\varphi(\underset{\sim}{Q})f.$$

Since $(H_o + I)\varphi(\underset{\sim}{Q})f$ defines a vector in $\mathcal{H}$, the remaining assertions of the corollary follow from Proposition 2.32. □

As in Proposition 2.32, the result of Proposition 2.34 can be generalized to functions φ which do not vanish at infinity if v is not singular at infinity. For this it is useful to introduce the following notations.

Definition 2.36 : Let Γ be a *bounded* closed set of Lebesgue measure zero in $\mathbb{R}^\nu$. Then we denote by C_Γ the set of all functions φ in $C^\infty(\mathbb{R}^\nu)$ such that φ, gradφ and $\Delta\varphi$ are in $L^\infty(\mathbb{R}^\nu)$ and φ vanishes in an open neighbourhood of Γ. Furthermore we denote by $\mathcal{V}_\Gamma$ the set of all measurable functions $v : \mathbb{R}^\nu \to \mathbb{R}$ which may be written as $v = v_1 + v_2$, with $v_1 \in L^\infty(\mathbb{R}^\nu)$ and $v_2\varphi \in L^p(\mathbb{R}^\nu)$ for some p satisfying $p \geq 2$, $p > \nu/2$ and for each $\varphi \in C_\Gamma$. (Thus $|v_2|^p$ is integrable in a neighbourhood of infinity as well as locally in $\mathbb{R}^\nu \backslash \Gamma$.)

It is not difficult to see that

$$v \in \mathcal{V}_\Gamma \iff v\varphi \in \mathcal{V}_\emptyset \quad \forall \varphi \in C_\Gamma. \tag{2.65}$$

Proposition 2.37 : Let $\mathcal{H} = L^2(\mathbb{R}^\nu)$, let Γ be a bounded closed set of measure zero and assume that $v \in \mathcal{V}_\Gamma$. Let H be a self-adjoint extension of $\hat{H}$. Then

(a) $H_o\varphi(\underset{\sim}{Q})(H - z)^{-1} \in \mathcal{B}(\mathcal{H}) \qquad \forall \varphi \in C_\Gamma, \ \forall z \in \rho(H)$,

(b) If $f \in D(\hat{H}^*)$ and $\varphi \in C_\Gamma$, then $\varphi(\underset{\sim}{Q})f \in D(H_o) \cap D(H) \cap D(V)$ and $H\varphi(\underset{\sim}{Q})f = H_o\varphi(\underset{\sim}{Q})f + V\varphi(\underset{\sim}{Q})f$.

Proof : Choose χ and φ_n as in part (ii) of the proof of Proposition 2.32. Then (2.59) holds if Φ is replaced by Φ_n. Taking the limit as $n \to \infty$, which exists by the arguments in part (ii) of the proof of Proposition 2.32, we see that (2.59) also holds for Φ satisfying the hypotheses of the present proposition. Once (2.59) is established, the remainder of the proof is exactly the same as that of Proposition 2.34 and of Corollary 2.35 ($\psi \in C_\Gamma$ is now such that $|\psi(\underline{x})| \leq 1$, $\psi(\underline{x})\varphi(\underline{x}) = \varphi(\underline{x})$ and $\psi\chi_n \in C_o^\infty(\mathbb{R}^\nu \backslash \Gamma)$). □

Finally we give an analogue of Lemma 2.31 :

<u>Lemma 2.38</u> : Let Γ, H and v be as in Proposition 2.37. Let $w \in V_\Gamma$. Then, for each $\varphi \in C_\Gamma$ and each $z \in \rho(H)$, the operator $W\varphi(\underline{Q})(H-z)^{-1}$ and the closure of $(H-z)^{-1}\varphi(\underline{Q})W$ are in $\mathcal{B}(\mathcal{H})$. Also, for each $m = 1,\ldots,\nu$, the operator $P_m\varphi(\underline{Q})(H-z)^{-1}$ and the closure of $(H-z)^{-1}\varphi(\underline{Q})P_m$ are in $\mathcal{B}(\mathcal{H})$.

<u>Proof</u> : (i) Let ψ be as indicated in the preceding proof. Then

$$W\varphi(\underline{Q})(H-z)^{-1} = [W\psi(\underline{Q})(H_o - i)^{-1}][(H_o - i)\varphi(\underline{Q})(H-z)^{-1}].$$

The first factor is in $\mathcal{B}(\mathcal{H})$ by Proposition 2.28, the second one by Proposition 2.37.

(ii) By Proposition 2.17 (b), $P_m(H_o - i)^{-1} \in \mathcal{B}(\mathcal{H})$. Hence

$$P_m\varphi(\underline{Q})(H-z)^{-1} = [P_m(H_o - i)^{-1}][(H_o - i)\varphi(\underline{Q})(H-z)^{-1}] \in \mathcal{B}(\mathcal{H}).$$

(iii) The proof of the other two assertions of the lemma uses the same reasoning as that employed in part (ii) of the proof of Lemma 2.31 and is omitted. □

<u>Problem 2.39</u> : Let $\gamma \in (-\infty, 3/4)$. Show that the differential equation $(-d^2/dr^2 + \gamma r^{-2} - i)f(r) = 0$ has a solution belonging to $L^2([0,\infty))$. Use this to prove that the operator $\hat{H} = -\Delta + \gamma|\underline{x}|^{-2}$ with domain $D(\hat{H}) = C_o^\infty(\mathbb{R}^3 \backslash \{\underline{0}\})$ is not essentially self-adjoint. Determine its self-adjoint extensions. (Hint : Write $\hat{H}$ in spherical polar coordinates. See also [AJS, Ch. 11].)

CHAPTER 3 : HILBERT-SCHMIDT AND COMPACT OPERATORS

In this chapter we introduce the Hilbert-Schmidt class and the class of compact operators and prove some of their properties which will be crucial for the further development of the theory.

3.1 Hilbert-Schmidt Operators

Let H be a separable Hilbert space, A an operator in $B(H)$. We associate with A a new norm $||A||_{HS}$, the Hilbert-Schmidt norm of A, as follows : let $\{e_i\}$ be an orthonormal basis of H. Then

$$||A||_{HS}^2 = \sum_i ||Ae_i||^2. \tag{3.1}$$

If $||A||_{HS} < \infty$, A is called a Hilbert-Schmidt operator. The set of all Hilbert-Schmidt operators on H will be denoted by $B_2(H)$ or simply by B_2.

Lemma 3.1 : The sum in (3.1) is independent of the chosen orthonormal basis $\{e_i\}$ of H (hence $||A||_{HS}$ depends only on A). Furthermore

$$||A^*||_{HS} = ||A||_{HS}. \tag{3.2}$$

Proof : Let $\{g_k\}$ be another orthonormal basis of H. Then

$$||Ag_k||^2 = \sum_i |(e_i, Ag_k)|^2,$$

hence

$$\sum_k ||Ag_k||^2 = \sum_{ki}\sum |(e_i, Ag_k)|^2 = \sum_{ik}\sum |(A^*e_i, g_k)|^2$$

$$= \sum_i ||A^*e_i||^2. \tag{3.3}$$

The interchanging of the order of summation is justified since all terms are positive. The r.h.s. of (3.3) is independent of the orthonormal basis $\{g_k\}$, hence the same for each basis $\{g_k\}$, which proves the first assertion.

(3.2) is an immediate consequence of (3.3). □

Proposition 3.2 : (a) $A \in \mathcal{B}_2 \iff A^* \in \mathcal{B}_2$.

(b) $\|A\| \le \|A\|_{HS}$. (3.4)

(c) If $A_1, A_2 \in \mathcal{B}_2$ and $\alpha \in \mathbb{C}$, then $(A_1 + \alpha A_2) \in \mathcal{B}_2$.

(d) If $A \in \mathcal{B}_2$ and $B \in \mathcal{B}(\mathcal{H})$, then $AB \in \mathcal{B}_2$ and $BA \in \mathcal{B}_2$, and

$$\|AB\|_{HS} \le \|B\| \, \|A\|_{HS}, \tag{3.5}$$

$$\|BA\|_{HS} \le \|B\| \, \|A\|_{HS}. \tag{3.6}$$

Proof : (a) follows from (3.3). For (b), let $f \neq \theta$. Choose an orthonormal basis $\{e_i\}$ of $\mathcal{H}$ such that $e_1 = \|f\|^{-1} f$. Then

$$\frac{\|Af\|^2}{\|f\|^2} = \|Ae_1\|^2 \le \sum_i \|Ae_i\|^2 = \|A\|_{HS}^2.$$

Since this holds for each $f \neq \theta$, we obtain

$$\|A\| = \sup_{f \neq \theta} \frac{\|Af\|}{\|f\|} \le \|A\|_{HS}.$$

(c) We have $\|f+g\|^2 \le 2\|f\|^2 + 2\|g\|^2$ by (1.12). This implies that

$$\|A_1 + \alpha A_2\|_{HS}^2 \le 2\|A_1\|_{HS}^2 + 2|\alpha|^2 \|A_2\|_{HS}^2 < \infty. \tag{3.7}$$

(d) Using (1.24) we get

$$\|BA\|_{HS}^2 = \sum_i \|BAe_i\|^2 \le \|B\|^2 \sum_i \|Ae_i\|^2 = \|B\|^2 \|A\|_{HS}^2,$$

which shows that $BA \in \mathcal{B}_2$ and proves (3.6). Similarly, since $B^* \in \mathcal{B}(H)$ and $A^* \in \mathcal{B}_2$, we have $B^*A^* \in \mathcal{B}_2$. Using (a), we see that $(B^*A^*)^* = AB \in \mathcal{B}_2$. Also $\|AB\|_{HS} = \|B^*A^*\|_{HS} \leq \|B^*\| \, \|A^*\|_{HS} = \|B\| \, \|A\|_{HS}$, which proves (3.5). □

<u>Proposition 3.3</u> : $\mathcal{B}_2(H)$ is a Hilbert space with respect to the scalar product

$$\langle A,B \rangle := \sum_i (Ae_i, Be_i), \tag{3.8}$$

where $\{e_i\}$ is an arbitrary orthonormal basis of H.

<u>Proof</u> : (i) Let $A,B \in \mathcal{B}_2$. We first show that $\langle A,B \rangle$ is finite. For this, we notice that

$$(\|Ae_i\| - \|Be_i\|)^2 = \|Ae_i\|^2 + \|Be_i\|^2 - 2\|Ae_i\| \, \|Be_i\| \geq 0, \tag{3.9}$$

hence, by (1.10) :

$$2|(Ae_i, Be_i)| \leq 2\|Ae_i\| \, \|Be_i\| \leq \|Ae_i\|^2 + \|Be_i\|^2.$$

It follows that

$$\sum_i |(Ae_i, Be_i)| \leq \frac{1}{2}\|A\|_{HS}^2 + \frac{1}{2}\|B\|_{HS}^2.$$

Thus the series in (3.8) is absolutely convergent, which proves our claim.

(ii) The polarization identity (1.14) implies that $\langle A,B \rangle =$ $= \frac{1}{4}(\|A+B\|_{HS}^2 - \|A-B\|_{HS}^2 + i\|A-iB\|_{HS}^2 - i\|A+iB\|_{HS}^2)$, which shows that $\langle A,B \rangle$ is independent of the basis $\{e_i\}$ occuring in (3.8).

(iii) By Proposition 3.2 (c), $\mathcal{B}_2$ is a linear vector space. By using the axiom (H2) for H, one easily sees that $\langle \cdot,\cdot \rangle$ satisfies (H2). We also have (denoting by Tr the trace in H) :

$$\|A\|^2_{HS} = \sum_i \|Ae_i\|^2 = \sum_i (Ae_i, Ae_i) = \langle A,A \rangle$$

$$= \sum_i (e_i, A^*Ae_i) \equiv \mathrm{Tr}\, A^*A. \tag{3.10}$$

(iv) It remains to show that $\mathcal{B}_2$ is complete with respect to $\|\cdot\|_{HS}$, i.e. to verify the axiom (H3). For this, suppose that $\{A_n\}$ is a Cauchy sequence in $\mathcal{B}_2$, i.e. such that $\|A_n - A_m\|_{HS} \to 0$ as $n,m \to \infty$. (3.4) then implies, for each $f \in \mathcal{H}$: $\|(A_n - A_m)f\| \leq \leq \|A_n - A_m\|_{HS}\|f\|$ as $n,m \to \infty$. Consequently $\{A_nf\}$ is Cauchy in $\mathcal{H}$. If we set $g = \text{s-lim}\, A_nf$, then the correspondence $f \mapsto g$ defines a linear operator A with domain $D(A) = \mathcal{H}$. We show that $\|A - A_n\|_{HS} \to 0$ as $n \to \infty$.

Given $\varepsilon > 0$, there is an integer N such that, for all n, $m > N$ and each $s = 1,2,\ldots$:

$$\sum_{i=1}^{s} \|(A_n - A_m)e_i\|^2 \leq \|A_n - A_m\|^2_{HS} < \varepsilon^2/4.$$

Then, by (1.12),

$$\sum_{i=1}^{s} \|(A_n - A)e_i\|^2 \leq 2\sum_{i=1}^{s} \|(A_n - A_m)e_i\|^2 + 2\sum_{i=1}^{s} \|(A_m - A)e_i\|^2$$
$$\leq \tfrac{1}{2}\varepsilon^2 + 2s \max_{1\leq i\leq s} \|(A_m - A)e_i\|^2 < \varepsilon^2,$$

since, for fixed s, one can choose m such that the second term is less than $\varepsilon^2/2$ ($\|A_me_i - Ae_i\| \to 0$ as $m \to \infty$). Therefore $\|A_n - A\|_{HS} < \varepsilon$ whenever $n > N$, which shows that $\|A_n - A\|_{HS} \to 0$ as $n \to \infty$.

Finally we have from (3.7) that

$$\|A\|^2 \leq \|A\|^2_{HS} \leq 2\|A - A_n\|^2_{HS} + 2\|A_n\|^2_{HS} < \infty,$$

proving that $A \in \mathcal{B}_2$. □

We next show that, if H is a function space, then Hilbert-Schmidt operators have a simple (and very useful) characterization as integral operators. So let $H = L^2(\Delta;d\mu)$ be the Hilbert space of all measurable functions from a set Δ to $\mathbb{C}$ that are square-integrable with respect to a measure $\mu(\cdot)$ on Δ (in our applications μ will be Lebesgue measure). An operator A in H is said to be an integral operator if there is a measurable function $a : \Delta \times \Delta \to \mathbb{C}$ such that

$$(Af)(s) = \int_\Delta a(s,s')f(s')d\mu(s') \quad (s \in \Delta) \tag{3.11}$$

for all $f \in D(A)$. a is called the kernel of A.

Definition : A function $a : \Delta \times \Delta \to \mathbb{C}$ is called a Hilbert-Schmidt kernel (with respect to the measure μ) if

$$M_a := \int_\Delta \int_\Delta d\mu(s)d\mu(s')|a(s,s')|^2 < \infty. \tag{3.12}$$

Proposition 3.4 : Let $H = L^2(\Delta;d\mu)$. Then $A \in B_2(H)$ if and only if A is an integral operator with Hilbert-Schmidt kernel. Furthermore

$$||A||^2_{HS} = M_a = \int_\Delta \int_\Delta d\mu(s)d\mu(s')|a(s,s')|^2. \tag{3.13}$$

Remark : In the proof of this result we shall use the fact that $L^2(\Delta \times \Delta;d\mu(s)d\mu(s')) = L^2(\Delta;d\mu(s)) \otimes L^2(\Delta;d\mu(s'))$. More precisely, we shall use the following fact : If $\{e_i(s)\}$ and $\{f_i(s)\}$ are two orthonormal bases of $L^2(\Delta;d\mu)$, then $\{e_i(s)f_j(s')\}^\infty_{i,j=1}$ is an orthonormal basis of $L^2(\Delta \times \Delta;d\mu(s)d\mu(s'))$. Also we refrain from specifying the measure $d\mu$ in the proof below.

Proof : (a) Let A be an integral operator with Hilbert-Schmidt kernel a(s,s').

(i) Let $\Delta_0 = \{s \in \Delta | \int_\Delta |a(s,s')|^2 d\mu(s') = \infty\}$. Then the measure of Δ_0 is zero, since $a(\cdot,\cdot) \in L^2(\Delta \times \Delta)$. Let $f \in L^2(\Delta)$ and $s \notin \Delta_0$. We view the integral in (3.11) as a scalar product in $L^2(\Delta)$ and apply to it the Schwarz inequality to obtain that

$$|(Af)(s)|^2 \leq \int_\Delta |a(s,s')|^2 d\mu(s') \cdot \int_\Delta |f(s')|^2 d\mu(s').$$

Integrating this inequality over s, we get $\|Af\|^2 \leq M_a \|f\|^2$. Hence $D(A) = H$ and $A \in B(H)$, with $\|A\| \leq M_a^{1/2}$.

(ii) Let $\{e_i\}$ be an orthonormal basis of H. For $A \in B(H)$, define $\alpha_{ik} = (e_i, Ae_k)$. Then $Ae_k = \sum_i \alpha_{ik} e_i$ and

$$\|A\|_{HS}^2 = \sum_k \|Ae_k\|^2 = \sum_{ik} |\alpha_{ik}|^2. \tag{3.14}$$

(iii) If $\{e_k(s)\}$ is an orthonormal basis of $L^2(\Delta)$, then the complex-conjugate functions $\{\overline{e_k(s)}\}$ also form an orthonormal basis. Hence the functions $\{e_i(s)\overline{e_k(s')}\}$ form an orthonormal basis of $L^2(\Delta \times \Delta)$. Since $a(\cdot,\cdot) \in L^2(\Delta \times \Delta)$, it may be developed with respect to this basis, i.e. one may write $a = \sum \beta_{ik} e_i \overline{e_k}$ and then has

$$M_a = \sum_{ik} |\beta_{ik}|^2. \tag{3.15}$$

Now

$$\alpha_{ik} = (e_i, Ae_k) = \int_\Delta \int_\Delta \overline{e_i(s)} a(s,s') e_k(s') d\mu(s) d\mu(s')$$

$$= (e_i \overline{e_k}, a)_{L^2(\Delta \times \Delta)} = \beta_{ik}. \tag{3.16}$$

(3.14)-(3.16) imply that

$$\|A\|_{HS}^2 = \sum_{ik} |\alpha_{ik}|^2 = \sum_{ik} |\beta_{ik}|^2 = M_a,$$

in particular that $A \in B_2$ if M_a is finite.

(b) Suppose $A \in B_2$. Let α_{ik} be as above and set

$$a_N(s,s') = \sum_{i,k=1}^{N} \alpha_{ik} e_i(s) \overline{e_k(s')}.$$

a_N is a finite linear combination of functions in $L^2(\Delta \times \Delta)$, hence

$a_N(\cdot,\cdot) \in L^2(\Delta \times \Delta)$. Let $M > N$. One then has :

$$\int_\Delta \int_\Delta |a_N(s,s') - a_M(s,s')|^2 d\mu(s) d\mu(s')$$

$$= \sum_{i=1}^{N} \sum_{k=N+1}^{M} |\alpha_{ik}|^2 + \sum_{i=N+1}^{M} \sum_{k=1}^{M} |\alpha_{ik}|^2$$

$$\leq \sum_{i=1}^{\infty} \sum_{k=N+1}^{\infty} |\alpha_{ik}|^2 + \sum_{i=N+1}^{\infty} \sum_{k=1}^{\infty} |\alpha_{ik}|^2 . \tag{3.17}$$

Since $A \in \mathcal{B}_2$, the double sum in (3.14) is convergent. Thus (3.17) implies that $\{a_N\}$ is a Cauchy sequence in $L^2(\Delta \times \Delta)$. We denote by $a(\cdot,\cdot)$ its limit function in $L^2(\Delta \times \Delta)$. By (a) above, the kernel $a(\cdot,\cdot)$ defines a Hilbert-Schmidt operator in $L^2(\Delta)$, which we denote by A_o.

It remains to show that $A_o = A$. Since both operators are in $\mathcal{B}(\mathcal{H})$, it suffices to check that $A_o e_k = Ae_k$ for each basis vector e_k. We know already that $Ae_k = \sum_i \alpha_{ik} e_i$. If we write $A_o e_k = \sum_i \alpha^o_{ik} e_i$, we have by (3.16) :

$$\alpha^o_{ik} = (e_i, A_o e_k) = (e_i \overline{e_k}, a)_{L^2(\Delta \times \Delta)}$$

$$= \lim_{N \to \infty} (e_i \overline{e_k}, a_N)_{L^2(\Delta \times \Delta)} = \alpha_{ik} . \qquad \square$$

Problem 3.5 : Show that the Hilbert space $\mathcal{B}_2(\mathcal{H})$ may be identified with $\mathcal{H} \otimes \mathcal{H}$, the tensor product of $\mathcal{H}$ with itself. (First identify $\mathcal{H}$ with ℓ^2 by fixing an orthonormal basis $\{e_i\}$ in $\mathcal{H}$.)

Some of the importance of Hilbert-Schmidt operators resides in the following facts :

(a) one may estimate the norm of certain operators by calculating or estimating an integral,

(b) Hilbert-Schmidt operators are often used in proving that certain operators are compact (see Section 3.2),

(c) scattering cross sections may be expressed in terms of Hilbert-Schmidt norms (see Chapter 6).

Proposition 3.6 : Let $\mathcal{H} = L^2(\mathbb{R}^\nu)$, $\underline{Q} = (Q_1,\ldots,Q_\nu)$ the position operator and $\underline{P} = (P_1,\ldots,P_\nu)$ the momentum operator. Let $\varphi,\psi:\mathbb{R}^\nu \to \mathbb{C}$, let $\psi(\underline{Q})$ be the multiplication operator by $\psi(\underline{x})$ in $L^2(\mathbb{R}^\nu)$ and $\varphi(\underline{P})$ the multiplication operator by $\varphi(\underline{k})$ in $\tilde{L}^2(\mathbb{R}^\nu)$. Then $\psi(\underline{Q})\varphi(\underline{P})$ and $\varphi(\underline{P})\psi(\underline{Q})$ are Hilbert-Schmidt operators if and only if either both functions φ and ψ belong to $L^2(\mathbb{R}^\nu)$ or one of them is zero almost everywhere, and one has

$$\|\psi(\underline{Q})\varphi(\underline{P})\|_{HS} = \|\varphi(\underline{P})\psi(\underline{Q})\|_{HS} = (2\pi)^{-\nu/2}\|\varphi\|_2\|\psi\|_2. \quad (3.18)$$

Proof : We give the proof for $\varphi(\underline{P})\psi(\underline{Q})$.

(i) First assume that $\varphi\in L^2(\mathbb{R}^\nu)$. For fixed $\underline{x}\in\mathbb{R}^\nu$, we then have

$$[\varphi(\underline{P})f](\underline{x}) = (2\pi)^{-\nu/2}\int d^\nu k e^{i\underline{k}\cdot\underline{x}}\varphi(\underline{k})\tilde{f}(\underline{k}).$$

The integral may be viewed as the scalar product between the functions $\exp(-i\underline{k}\cdot\underline{x})\bar{\varphi}(\underline{k})$ and $\tilde{f}(\underline{k})$. By the unitarity of the Fourier transformation, this is equal to the scalar product between the inverse Fourier transforms of these two functions. Now

$$[F^{-1}\{\bar{\varphi}(\underline{k})e^{-i\underline{k}\cdot\underline{x}}\}](\underline{y}) = (2\pi)^{-\nu/2}\int d^\nu k e^{i\underline{k}\cdot\underline{y}}e^{-i\underline{k}\cdot\underline{x}}\bar{\varphi}(\underline{k})$$

$$= \overline{\tilde{\varphi}(\underline{y}-\underline{x})}.$$

Hence $[\varphi(\underline{P})f](\underline{x}) = (2\pi)^{-\nu/2}\int d^\nu y\tilde{\varphi}(\underline{y}-\underline{x})f(\underline{y})$. (3.19)

This shows that $\varphi(\underline{P})$ is an integral operator in $L^2(\mathbb{R}^\nu)$ with kernel $a(\underline{x},\underline{x}') = (2\pi)^{-\nu/2}\tilde{\varphi}(\underline{x}'-\underline{x})$.

(ii) $\varphi(\underline{P})\psi(\underline{Q})$ is an integral operator with kernel $(2\pi)^{-\nu/2}\tilde{\varphi}(\underline{x}'-\underline{x})\psi(\underline{x}')$. Thus, by Proposition 3.4 :

$$\|\varphi(\underline{P})\psi(\underline{Q})\|_{HS}^2 = (2\pi)^{-\nu}\int\int d^\nu x d^\nu x' |\tilde{\varphi}(\underline{x}' - \underline{x})|^2 |\psi(\underline{x}')|^2$$

$$= (2\pi)^{-\nu}\int d^\nu k|\tilde{\varphi}(\underline{k})|^2 \int d^\nu x' |\psi(\underline{x}')|^2 = (2\pi)^{-\nu}\|\varphi\|_2^2 \|\psi\|_2^2 ,$$

where we have made the change of variables $\underline{x} \mapsto \underline{k} = \underline{x}' - \underline{x}$ and used (1.18). The preceding identity, which is just (3.18), proves that $\varphi(\underline{P})\psi(\underline{Q})\in\mathcal{B}_2$ if $\varphi,\psi\in L^2(\mathbb{R}^\nu)$ and that $\varphi(\underline{P})\psi(\underline{Q})\notin\mathcal{B}_2$ if $\varphi\in L^2(\mathbb{R}^\nu)$ but $\psi\notin L^2(\mathbb{R}^\nu)$.

(iii) If $\varphi\notin L^2(\mathbb{R}^\nu)$, define $\varphi_n \in L^2(\mathbb{R}^\nu)$ for $n > 0$ by

$$\varphi_n(\underline{k}) = \begin{cases} \varphi(\underline{k}) & \text{if } |\underline{k}| \le n \text{ and } |\varphi(\underline{k})| \le n \\ n & \text{if } |\underline{k}| \le n \text{ and } |\varphi(\underline{k})| > n \\ 0 & \text{if } |\underline{k}| > n. \end{cases}$$

Then $|\varphi_n(\underline{k})| \le |\varphi(\underline{k})|$ for all $\underline{k}\in\mathbb{R}^\nu$, hence $\|\varphi(\underline{P})g\| \ge \|\varphi_n(\underline{P})g\|$ for each $g\in L^2(\mathbb{R}^\nu)$. This implies that

$$\|\varphi(\underline{P})\psi(\underline{Q})\|_{HS}^2 \ge \|\varphi_n(\underline{P})\psi(\underline{Q})\|_{HS}^2 = (2\pi)^{-\nu}\|\varphi_n\|_2^2 \|\psi\|_2^2.$$

Since $\varphi\notin L^2(\mathbb{R}^\nu)$ and $|\varphi_n(\underline{k})| \to |\varphi(\underline{k})|$ for each $\underline{k}\in\mathbb{R}^\nu$, we have $\|\varphi_n\|_2^2 \to \infty$ as $n \to \infty$ by the monotone convergence theorem [R]. Hence $\|\varphi(\underline{P})\psi(\underline{Q})\|_{HS} = \infty$, i.e. $\varphi(\underline{P})\psi(\underline{Q})\notin\mathcal{B}_2$, unless $\|\psi\|_2 = 0$. □

3.2 Compact Operators

If $A\in\mathcal{B}_2$ and $\{e_k\}$ is an orthonormal basis of $\mathcal{H}$, we may, for each integer $N < \infty$, define an operator A_N by

$$A_N\left[\sum_{k=1}^{\infty} \alpha_k e_k\right] = \sum_{k=1}^{N} \alpha_k A e_k .$$

We see in particular that $A_N e_k = Ae_k$ if $k \le N$ and $A_N e_k = \theta$ if $k > N$. The range of A_N is a finite-dimensional subspace, the subspace spanned by $Ae_1,\ldots,Ae_N$. A_N is called a finite rank operator.

It may be written as $A_N f = \sum_{k=1}^{N}(e_k,f)Ae_k$ $(f \in \mathcal{H})$. More generally, a finite rank operator is defined to be an operator of the form

$$Tf = \sum_{k=1}^{N} (g_k,f)h_k \ , \tag{3.20}$$

where $N < \infty$ and $g_1,\ldots,g_N$, $h_1,\ldots,h_N$ are 2N arbitrary vectors in $\mathcal{H}$.

In the example given above, one has $\|A - A_N\|_{HS}^2 = \sum_{k=N+1}^{\infty}\|Ae_k\|^2$, which converges to zero as $N \to \infty$, since $A \in \mathcal{B}_2$. Hence each Hilbert-Schmidt operator is the limit in Hilbert-Schmidt norm of a sequence of finite rank operators (and clearly each finite rank operator is in $\mathcal{B}_2$). One obtains a larger class of bounded operators by considering all *uniform* limits of sequences $\{T_N\}$ of finite rank operators.

Definition : An operator A in $\mathcal{B}(\mathcal{H})$ is compact if there is a sequence $\{T_N\}$ of finite rank operators such that $\|A - T_N\| \to 0$ as $N \to \infty$. The set of all compact operators acting in $\mathcal{H}$ will be denoted by $\mathcal{B}_\infty(\mathcal{H})$ or simply by $\mathcal{B}_\infty$.

Lemma 3.7 : Let T be a finite rank operator. Then

(a) T^* is a finite rank operator.

(b) If $B \in \mathcal{B}(\mathcal{H})$, then BT and TB are finite rank operators.

(c) If T_1 is a finite rank operator, then so is $T + \alpha T_1$ $(\alpha \in \mathbb{C})$.

Proof : (c) is evident. (b) is also straightforward, since, if T is given by (3.20), then

$$BTf = \sum_{k=1}^{N} (g_k,f)Bh_k \ , \quad TBf = \sum_{k=1}^{N} (B^*g_k,f)h_k .$$

(a) One has for $f,g \in \mathcal{H}$:

$$(f,T^*g) = (Tf,g) = \sum_{k=1}^{N} \overline{(g_k,f)}(h_k,g) = (f, \sum_{k=1}^{N} (h_k,g)g_k).$$

Hence

$$T^*g = \sum_{k=1}^{N} (h_k,g)g_k \ , \qquad (3.21)$$

i.e. T^* is a finite rank operator. □

Proposition 3.8 :

(a) $A\in\mathcal{B}_\infty$ if and only if $A^*\in\mathcal{B}_\infty$.

(b) If $A\in\mathcal{B}_\infty$ and $B\in\mathcal{B}(\mathcal{H})$, then $AB\in\mathcal{B}_\infty$ and $BA\in\mathcal{B}_\infty$.

(c) If $A_1,A_2\in\mathcal{B}_\infty$ and $\alpha\in\mathbb{C}$, then $(A_1 + \alpha A_2)\in\mathcal{B}_\infty$.

(d) If $A_n\in\mathcal{B}_\infty$ $(n = 1,2,\ldots)$, $A\in\mathcal{B}(\mathcal{H})$ and $||A - A_n|| \to 0$ as $n \to \infty$, then $A\in\mathcal{B}_\infty$.

(e) $\mathcal{B}_2\subseteq\mathcal{B}_\infty$, i.e. each Hilbert-Schmidt operator is compact.

Proof : (a) Let $\{T_N\}$ be a sequence of finite rank operators such that $||A - T_N|| \to 0$. Then T_N^* are finite rank operators, and $||A^* - T_N^*|| = ||A - T_N|| \to 0$. Hence $A\in\mathcal{B}_\infty$ implies $A^*\in\mathcal{B}_\infty$. Similarly, $A^*\in\mathcal{B}_\infty$ implies $A^{**}\in\mathcal{B}_\infty$, hence $A\in\mathcal{B}_\infty$ since $A^{**} = A$.

(b) BT_N and T_NB are finite rank operators by Lemma 3.7 (b). Hence $||BA - BT_N|| \le ||B||\,||A - T_N|| \to 0$, i.e. $BA\in\mathcal{B}_\infty$. Similarly $||AB - T_NB|| \le ||A - T_N||\,||B|| \to 0$, hence $AB\in\mathcal{B}_\infty$.

(c) This follows from Lemma 3.7 (c) and (1.26) :

$$||A_1 + \alpha A_2 - T_N^{(1)} - \alpha T_N^{(2)}|| \le ||A_1 - T_N^{(1)}|| + |\alpha|\,||A_2 - T_N^{(2)}|| \to 0.$$

(d) For each n, choose a finite rank operator $T_{N(n)}$ such that $||A_n - T_{N(n)}|| < 1/n$. Given $\varepsilon > 0$, there is a number n_0 such that, for all $n \ge n_0$, $||A_n - T_{N(n)}|| < \varepsilon/2$ and $||A - A_n|| < \varepsilon/2$. Hence, for $n \ge n_0$,

$$||A - T_{N(n)}|| \le ||A - A_n|| + ||A_n - T_{N(n)}|| < \varepsilon.$$

This shows that $||A - T_{N(n)}|| \to 0$ as $n \to \infty$. Hence $A \in \mathcal{B}_\infty$.

(e) We have seen at the beginning of this section that, if $A \in \mathcal{B}_2$ there is a sequence $\{A_N\}$ of finite rank operators such that $||A - A_N||_{HS} \to 0$. It follows from (3.4) that $||A - A_N|| \le ||A - A_N||_{HS} \to 0$, so that $A \in \mathcal{B}_\infty$. □

<u>Proposition 3.9</u> : Let $A \in \mathcal{B}_\infty$ and let $\{f_n\}$ be a sequence of vectors converging weakly to zero, i.e. w-lim $f_n = \theta$. Then s-lim $Af_n = \theta$ as $n \to \infty$. (In other words, a compact operator maps each weakly convergent sequence into a strongly convergent sequence).

<u>Proof</u> : We use the fact that each weakly convergent sequence $\{f_n\}$ is bounded, i.e. that there is a number $M < \infty$ such that $||f_n|| \le M$ for all n. We do not prove this result; however, in our applications of Proposition 3.9, the existence of such a bound is always immediate. (Reference : [AG, no. 23]).

Let $\varepsilon > 0$. Let T be a finite rank operator, of the form (3.20) such that $||A - T|| < \varepsilon/2M$. Then

$$||Af_n|| \le ||(A - T)f_n|| + ||Tf_n|| < \varepsilon/2 + \sum_{k=1}^{N} |(g_k, f_n)|\,||h_k||.$$

Since w-lim $f_n = \theta$, there is a number n_o such that $|(g_k, f_n)|\,||h_k|| < \varepsilon/2N$ for each $k = 1, \ldots, N$ and all $n \ge n_o$. Hence $||Af_n|| < \varepsilon$ for all $n \ge n_o$, which proves that $||Af_n|| \to 0$. □

<u>Problem 3.10</u> : A projection E is compact if and only if its range $\mathcal{M}(E)$ is a finite-dimensional subspace.

<u>Proposition 3.11</u> : Let $B_n, B \in \mathcal{B}(\mathcal{H})$ be such that s-lim $B_n = B$ as $n \to \infty$.

(a) If $A \in \mathcal{B}_\infty$, then $||B_nA - BA|| \to 0$ and $||AB_n^* - AB^*|| \to 0$, in other words u-lim $B_nA = BA$ and u-lim $AB_n^* = AB^*$ as $n \to \infty$.

(b) If $A \in \mathcal{B}_2$, then $||B_nA - BA||_{HS} \to 0$ and $||AB_n^* - AB^*||_{HS} \to 0$ as $n \to \infty$.

Proof : (i) Let T be a finite rank operator of the form (3.20). Let $\{e_1,\ldots,e_m\}$ ($m < \infty$) be an orthonormal basis of the subspace spanned by $g_1,\ldots,g_N$. Let $C_n = B_n - B$. Then, by (1.13),

$$\|C_n T\|_{HS}^2 = \sum_{i=1}^{m} \|C_n Te_i\|^2 = \sum_{i=1}^{m} \|\sum_{k=1}^{N} (g_k,e_i)C_n h_k\|^2$$

$$\leq N \sum_{i=1}^{m} \sum_{k=1}^{N} |(g_k,e_i)|^2 \|C_n h_k\|^2 .$$

Since s-lim $C_n = 0$ and the preceding sum is finite, we have

$$\|C_n T\|_{HS} = \|(B_n - B)T\|_{HS} \to 0 \quad \text{as} \quad n \to \infty. \tag{3.22}$$

(ii) If $A \in \mathcal{B}_\infty$, there is a sequence $\{T_N\}$ of finite rank operators such that $\|A - T_N\| \to 0$ as $N \to \infty$. Then

$$\|(B_n - B)A\| \leq (\|B_n\| + \|B\|)\|A - T_N\| + \|(B_n - B)T_N\|. \tag{3.23}$$

As in the preceding proof, there is a $M < \infty$ such that $\|B_n\| \leq M$ for all n. Hence

$$\|(B_n - B)A\| \leq (M + \|B\|)\|A - T_N\| + \|(B_n - B)T_N\|_{HS}.$$

Given $\varepsilon > 0$, first choose N such that $\|A - T_N\| < \varepsilon(2M + 2\|B\|)^{-1}$ and then n_o such that $\|(B_n - B)T_N\|_{HS} < \varepsilon/2$ for $n \geq n_o$. Then $\|(B_n - B)A\| < \varepsilon$ for $n \geq n_o$, i.e. u-lim $B_n A = BA$. This proves the first assertion in (a). The second one follows from this by taking the adjoint, using Proposition 3.8 (a).

(iii) The proof of (b) is very similar. It suffices to use, instead of (3.23), the inequality

$$\|(B_n - B)A\|_{HS} \leq (\|B_n\| + \|B\|)\|A - T_N\|_{HS} + \|(B_n - B)T_N\|_{HS} ,$$

where $\{T_N\}$ is a sequence of finite rank operators converging to A in Hilbert-Schmidt norm. □

In the next proposition we show that a self-adjoint compact operator has very special spectral properties.

Proposition 3.12 : Let A be a self-adjoint compact operator. Then

(a) A has pure point spectrum, i.e. $\mathcal{H}_c(A) = \{\theta\}$.

(b) Given any $\varepsilon > 0$, there is at most a finite number of eigenvalues λ of A such that $|\lambda| \geq \varepsilon$.

(c) Each non-zero eigenvalue of A has finite multiplicity, i.e. the corresponding eigensubspace is finite-dimensional.

Thus, the spectrum of a self-adjoint compact operator consists only of eigenvalues plus (if $\dim \mathcal{H} = \infty$) the point $\lambda = 0$. The only possible accumulation point of these eigenvalues is $\lambda = 0$, and each non-zero eigenvalue is of finite multiplicity. If $\mathcal{H}$ is infinite-dimensional, $\lambda = 0$ must be either an accumulation point of eigenvalues or an eigenvalue of infinite multiplicity. If $\{\lambda_i\}$ is an enumeration of the non-zero eigenvalues of A, $\{M_i\}$ the corresponding eigensubspaces, then

$$\mathcal{H} = \bigoplus_i M_i \oplus N(A). \tag{3.24}$$

Proof : The proof of part (a) will be given at the end of Section 4.2. For (b) and (c), let $\{e_n\}$ be an infinite orthonormal sequence of eigenvectors of A, i.e. $Ae_n = \lambda_n e_n$ (we assume $\dim \mathcal{H} = \infty$, otherwise (b) and (c) are evident). By Example 1.2, $\text{w-lim}\, e_n = \theta$ as $n \to \infty$, hence by Proposition 3.9,

$$\lim_{n\to\infty} ||Ae_n|| = \lim_{n\to\infty} |\lambda_n| = 0.$$

Hence an infinite sequence of different eigenvalues (if it exists) must accumulate at $\lambda = 0$, which proves (b). Also it is impossible to have an infinite sequence of mutually orthogonal eigenvectors corresponding to the same fixed eigenvalue λ, unless $\lambda = 0$. This proves (c). □

Finally we prove the compactness of some particular operators in $L^2(\mathbb{R}^\nu)$. This result will be used in Section 5.4.

Lemma 3.13 : Let $H = L^2(\mathbb{R}^\nu)$, $2 \le p < \infty$ and $\varphi, \psi \in L^p(\mathbb{R}^\nu)$. Then the operators $A_{\varphi\psi} = \varphi(\underline{P})\psi(\underline{Q})$ and $B_{\varphi\psi} = \psi(\underline{Q})\varphi(\underline{P})$ are compact.

Proof : (i) Let χ_R be the characteristic function of the ball $B_R := \{\underline{\xi} \in \mathbb{R}^\nu \mid |\underline{\xi}| \le R\}$, i.e. $\chi_R(\underline{\xi}) = 1$ if $|\underline{\xi}| \le R$ and $\chi_R(\underline{\xi}) = 0$ if $|\underline{\xi}| > R$. By Lemma A.1 (a), we have $\chi_R(\underline{x})\psi(\underline{x}) \in L^2(\mathbb{R}^\nu)$ and $\chi_R(\underline{k})\varphi(\underline{k}) \in L^2(\mathbb{R}^\nu)$. Thus, if we set $C_R = \chi_R(\underline{P})\varphi(\underline{P})$, $D_R = \chi_R(\underline{Q})\psi(\underline{Q})$, we have $C_R D_R \in \mathcal{B}_2$ by Proposition 3.6.

(ii) Since $\varphi, \psi \in L^p(\mathbb{R}^\nu)$, we have $\|[1 - \chi_R(\underline{x})]\psi(\underline{x})\|_p \to 0$ and $\|[1 - \chi_R(\underline{k})]\varphi(\underline{k})\|_p \to 0$ as $R \to \infty$. Hence we obtain from Lemma 2.27 that

$$\|\varphi(\underline{P})\psi(\underline{Q}) - C_R D_R\| \le \|[\varphi(\underline{P}) - C_R]\psi(\underline{Q})\| + \|C_R[\psi(\underline{Q}) - D_R]\|$$

$$\le \|[1 - \chi_R(\underline{k})]\varphi(\underline{k})\|_p \|\psi\|_p + \|\varphi\|_p \|[1 - \chi_R(\underline{x})]\psi(\underline{x})\|_p \to 0$$

as $R \to \infty$. Thus $\varphi(\underline{P})\psi(\underline{Q})$ is compact as the uniform limit of the sequence $\{C_R D_R\}_{R=1}^{\infty}$ of Hilbert-Schmidt operators, cf. Proposition 3.8 (d). Similarly one shows that $\psi(\underline{Q})\varphi(\underline{P}) \in \mathcal{B}_\infty$. □

Corollary 3.14 : Let $H = L^2(\mathbb{R}^\nu)$. Let H_o be the Schrödinger free Hamiltonian (2.28) and let $z \in \rho(H_o)$.

(a) If $(I + |\underline{Q}|)^{-\kappa}$ is the multiplication operator by $(1 + |\underline{x}|)^{-\kappa}$ and $\kappa > 0$, then $(I + |\underline{Q}|)^{-\kappa}(H_o - z)^{-1} \in \mathcal{B}_\infty$ and $(I + |\underline{Q}|)^{-\kappa} P_m (H_o - z)^{-1} \in \mathcal{B}_\infty$ for $m = 1, \ldots, \nu$.

(b) For each bounded function $\varphi \in L^2(\mathbb{R}^\nu)$, one has $\varphi(\underline{Q})(H_o - z)^{-1} \in \mathcal{B}_\infty$ and $\varphi(\underline{Q}) P_m (H_o - z)^{-1} \in \mathcal{B}_\infty$ $(m = 1, \ldots, \nu)$. In particular this holds for each $\varphi \in C_o^\infty(\mathbb{R}^\nu)$.

Proof : It suffices to remark that the functions $\underline{x} \mapsto (1 + |\underline{x}|)^{-\kappa}$, $\underline{x} \mapsto \varphi(\underline{x})$, $\underline{k} \mapsto (\underline{k}^2 - z)^{-1}$ and $\underline{k} \mapsto k_m(\underline{k}^2 - z)^{-1}$ belong to $L^p(\mathbb{R}^\nu)$ for all $p \in (p_0, \infty)$, where p_0 is finite and depends on ν and κ. (Notice that $\varphi \in L^2(\mathbb{R}^\nu) \cap L^\infty(\mathbb{R}^\nu)$, hence $\varphi \in L^q(\mathbb{R}^\nu)$ for each $q \in [2, \infty)$ by Lemma A.1 (b)). □

CHAPTER 4 : EVOLUTION GROUPS

An evolution group in quantum mechanics is a strongly continuous one-parameter group of unitary operators in a Hilbert space $\mathcal{H}$. (Strictly speaking, one uses the term "evolution group" only for a group giving the time evolution of a system, in which case the infinitesimal generator must be the Hamiltonian of the system. Since for our applications we have precisely this situation in mind, we use the term "evolution group", with some abuse of terminology, for an arbitrary strongly continuous unitary one-parameter group.) In Section 4.1 we establish the fact that there is a one-to-one correspondence between the set of evolution groups and the set of self-adjoint operators on $\mathcal{H}$. In Section 4.2 we define functions of a self-adjoint operator in terms of the associated evolution group and prove some simple results about this functional calculus. In Section 4.3 we show that vectors in the subspace of continuity $\mathcal{H}_c(A)$ of a self-adjoint operator A have certain ergodic properties, and in Section 4.4 we collect some results about the evolution group associated with the Schrödinger free Hamiltonian H_o.

4.1 Evolution Groups and Their Infinitesimal Generators

A strongly continuous unitary one-parameter group (or in short an evolution group) is a mapping $U:\mathbb{R} \to \mathcal{B}(\mathcal{H})$ having the following three properties :

(E1) Unitarity :

$$U_t^* U_t = U_t U_t^* = I \quad \forall\, t \in \mathbb{R}. \tag{4.1}$$

(E2) Strong continuity :

$$\text{s-}\lim_{\tau\to 0} (U_{t+\tau} - U_t) = 0 \qquad \forall\, t\in\mathbb{R}. \tag{4.2}$$

(E3) Group property :

$$U_t U_s = U_s U_t = U_{s+t} \qquad \forall\, s,t\in\mathbb{R}, \tag{4.3}$$

$$U_o = I. \tag{4.4}$$

Notice that, in order to simplify the notations, we have written U_t rather than U(t) for the value of U at the point t. Since $\mathcal{H}$ is separable, it would suffice to assume in (E2) only weak continuity or even only weak measurability (cf. [HP, Sec. 3.5 and 10.2]).

We notice that (E1) implies that $U_t^* = (U_t)^{-1}$, whereas (E3) gives $(U_t)^{-1} = U_{-t}$. Hence we have

$$U_t^* = U_{-t} \qquad \forall\, t\in\mathbb{R}. \tag{4.5}$$

The following theorem associates with each evolution group a self-adjoint operator.

Proposition 4.1 (Stone's Theorem) : Let $\{U_t\}$ be an evolution group. Define A to be the following linear operator, called the infinitesimal generator of $\{U_t\}$:

$$D(A) = \{f\in\mathcal{H} \mid \text{s-}\lim_{\tau\to 0} \tau^{-1}(U_\tau - I)f \ \text{ exists}\}, \tag{4.6}$$

$$Af = \text{s-}\lim_{\tau\to 0} i\tau^{-1}(U_\tau - I)f \quad \text{for} \quad f\in D(A). \tag{4.7}$$

Then D(A) is dense in $\mathcal{H}$ and A is self-adjoint.

Proof : (i) Let $z = x + iy$ with $x,y\in\mathbb{R}$ and $y > 0$. Let

$$R_z := i\int_0^\infty e^{izs} U_s\, ds. \tag{4.8}$$

This integral exists by Proposition 1.24 (d), since the integrand is strongly continuous and

$$\int_0^\infty \|e^{izs}U_s\| ds \le \int_0^\infty e^{-ys} ds = y^{-1} < \infty. \tag{4.9}$$

(4.9) and (1.63) imply that $R_z \in \mathcal{B}(\mathcal{H})$ and

$$\|R_z\| \le y^{-1}. \tag{4.10}$$

(ii) We now show that $\mathcal{R}(R_z)$ is dense in $\mathcal{H}$. For this we must prove that, if $(g,R_z f) = 0$ for each $f \in \mathcal{H}$, then $g = \theta$. Since

$$U_t R_z = R_z U_t \quad \text{for each} \quad t \in \mathbb{R}, \tag{4.11}$$

one has $(g,U_t R_z f) = (g,R_z(U_t f)) = 0$ for all $t \in \mathbb{R}$ and each $f \in \mathcal{H}$. Hence, inserting the definition (4.8) :

$$0 = \int_0^\infty e^{izs}(g,U_{t+s}f)ds = e^{-izt}\int_t^\infty e^{iz\sigma}(g,U_\sigma f)d\sigma.$$

By differentiating with respect to t, one obtains

$$0 = \frac{d}{dt}\int_t^\infty e^{iz\sigma}(g,U_\sigma f) = -e^{izt}(g,U_t f).$$

Hence $(g,U_t f) = 0$ for each t, in particular for $t = 0$. Thus $(g,f) = 0$ for each $f \in \mathcal{H}$, hence $g = \theta$.

(iii) $$i\tau^{-1}(U_\tau - I)R_z f = -\tau^{-1}\int_0^\infty e^{izs}U_{s+\tau}fds + \tau^{-1}\int_0^\infty e^{izs}U_s fds$$

$$= -\tau^{-1}(e^{-iz\tau} - 1)\int_\tau^\infty e^{iz\sigma}U_\sigma fd\sigma + \tau^{-1}\int_0^\tau e^{izs}U_s fds$$

$$= \frac{i}{\tau}(e^{-iz\tau} - 1)R_z f + \frac{1}{\tau}(e^{-iz\tau} - 1)\int_0^\tau e^{iz\sigma}U_\sigma fd\sigma + \frac{i}{\tau}\int_0^\tau e^{izs}U_s fds.$$

As $\tau \to 0$, the r.h.s. of this equation converges strongly to $zR_z f - iz\theta + f$. This implies that (a) $R_z f \in D(A)$ $\forall$ $f \in \mathcal{H}$ and (b) $AR_z f = zR_z f + f$, i.e.

$$(A - z)R_z f = f \quad \forall f \in \mathcal{H}. \tag{4.12}$$

(a) implies that D(A) is dense in H, since $R(R_z)$ is dense by (ii). By setting z = i in (4.12) we obtain that $R(A - iI) = H$.

(iv) By setting, for $z = x + iy$ with $y < 0$:

$$R_z = -i\int_{-\infty}^{0} e^{izs} U_s ds, \tag{4.13}$$

one obtains in the same way that (4.12) also holds for Imz < 0. In particular, for z = -i : $R(A+iI) = H$.

(v) To conclude that A = A*, it remains to show that A is symmetric (cf. Proposition 2.2). For this, let f,g∈D(A). Then

$$(Af,g) = \lim_{\tau\to 0}(i\tau^{-1}(U_\tau - I)f,g) = \lim_{\tau\to 0}(f,-i\tau^{-1}(U_\tau^* - I)g).$$

The symmetry of A now follows by using (4.5) and setting τ = -t :

$$(Af,g) = \lim_{t\to 0}(f,it^{-1}(U_t - I)g) = (f,Ag). \qquad \square$$

Corollary 4.2 : The resolvent of the infinitesimal generator A is related to the associated evolution group $\{U_t\}$ as follows :

$$\text{if } \operatorname{Im} z > 0 : (A - zI)^{-1} = i\int_0^{\infty} e^{izs} U_s ds, \tag{4.14}$$

$$\text{if } \operatorname{Im} z < 0 : (A - zI)^{-1} = -i\int_{-\infty}^{0} e^{izs} U_s ds. \tag{4.15}$$

Proof : Let e.g. Im z > 0. Since $(A - zI)^{-1} \in B(H)$ by Proposition 2.8, we obtain upon multiplying (4.12) by $(A - z)^{-1}$ that $R_z f = (A-z)^{-1}f$, which, by virtue of (4.8), is the identity (4.14). (4.15) is obtained similarly. □

Corollary 4.3 : Let $\{U_t\}$ and A be as in Proposition 4.1. Then U_t maps D(A) into itself, and

$$U_t Af = AU_t f \qquad \forall f\in D(A),\ \forall t\in \mathbb{R}. \tag{4.16}$$

Proof : By (4.3),

$$U_t \frac{i}{\tau}(U_\tau - I)f = \frac{i}{\tau}(U_\tau - I)U_t f.$$

If $f \in D(A)$, the left-hand side converges strongly to $U_t Af$ as $\tau \to 0$. Hence the right-hand side is also strongly convergent, which means that $U_t f \in D(A)$ and that the limit of the r.h.s. is $AU_t f$. □

A continuous symmetry group of a quantum-mechanical system must be represented in the relevant Hilbert space by a group of *unitary* operators depending on one or several parameters (for example the rotation group in $\mathbb{R}^3$ is a 3-parameter group). Stone's theorem shows that the infinitesimal generator(s) must be *self-adjoint*. In particular, if $\{U_t\}$ describes the time evolution of the system, its infinitesimal generator is the Hamiltonian of the system, which must be self-adjoint. This shows the necessity of studying self-adjointness properties of Schrödinger and Dirac operators.

Stone's theorem has a converse : given a self-adjoint operator A, there is a unique evolution group such that its infinitesimal generator is A. Formally we see from (4.16) and (4.7) that $AU_t = i dU_t/dt$, so that one should have $U_t = \exp(-iAt)$. If A is bounded, the exponential function may be defined by the usual power series $\exp(\alpha A) = \sum_{n=0}^{\infty} \alpha^n A^n/n!$, which is then convergent in the operator norm. If A is unbounded, A^n is still densely defined, since $D(A^n) = (A - i)^{-1} D(A^{n-1})$ and the image of a dense set under $(A - i)^{-1}$ is again dense. However the sets $D(A^n)$ are shrinking as n grows, and it is not so simple to construct a common dense domain. In this case it is convenient to use a different definition of the exponential function, which we shall do in (4.23) below to prove the converse of Stone's theorem.

Proposition 4.4 : Let A be a self-adjoint operator in H. Then there is a unique evolution group $\{U_t\}$ such that A is the infinitesimal generator of $\{U_t\}$.

Proof : (i) We set, for n = 1,2,...

$$U_{t,n} := (\frac{it}{n}A + I)^{-n} = \begin{cases} I & \text{if} \quad t = 0 \\ \left[\frac{n}{it}(A - \frac{in}{t})^{-1}\right]^n & \text{if} \quad t \neq 0. \end{cases} \tag{4.17}$$

Since, by (2.14),

$$\|(\frac{it}{n}A + I)^{-1}\| = \frac{n}{|t|}\|(A - \frac{in}{t})^{-1}\| \leq \frac{n}{|t|}\frac{|t|}{n} = 1, \tag{4.18}$$

we have $U_{t,n} \in \mathcal{B}(\mathcal{H})$ and

$$\|U_{t,n}\| \leq 1. \tag{4.19}$$

$t \mapsto U_{t,n}$ is strongly continuous. In fact, for $t \neq 0$, this function is even uniformly differentiable, by (2.18), and

$$\frac{dU_{t,n}}{dt} = -iA(\frac{it}{n}A + I)^{-n-1}. \tag{4.20}$$

At t = 0 we have for $f \in D(A)$

$$\|(\frac{it}{n}A + I)^{-1}f - f\| = \|\frac{it}{n}(\frac{it}{n}A + I)^{-1}Af\|$$

$$\leq \frac{|t|}{n}\|(\frac{it}{n}A + I)^{-1}\|\,\|Af\| \leq \frac{|t|}{n}\|Af\|, \tag{4.21}$$

which converges to zero as $t \to 0$. By Lemma 1.8, $\text{s-lim}(itn^{-1}A + I)^{-1} = I$ as $t \to 0$, and the strong continuity of $U_{t,n}$ at t = 0 follows from Lemma 1.9.

(ii) Let $f \in D(A^2)$. Then, by the results of (i) above, for t > 0 :

$$\|U_{t,n}f - U_{t,m}f\| = \|\text{s-lim}_{\delta \to +0} \int_\delta^{t-\delta} \frac{d}{ds} U_{t-s,m} U_{s,n} f ds\|$$

$$= \|\text{s-lim}_{\delta \to +0} \int_\delta^{t-\delta} ds (\frac{t-s}{m} - \frac{s}{n}) \left[\frac{i(t-s)}{m}A + I\right]^{-m-1} \cdot \left[\frac{is}{n}A + I\right]^{-n-1} A^2 f\|.$$

Since the integrand is strongly continuous, we may set $\delta = 0$ in the boundaries of the integral. Hence, by (1.57) and (4.18),

$$\|U_{t,n}f - U_{t,m}f\| \le \int_0^t \left(\frac{t-s}{m} + \frac{s}{n}\right) \|A^2 f\| ds = \frac{1}{2} t^2 \left(\frac{1}{m} + \frac{1}{n}\right) \|A^2 f\| . \quad (4.22)$$

Hence $\{U_{t,n}f\}$ is strongly Cauchy. Since $D(A^2)$ is dense in H and $\|U_{t,n}\| \le 1$, we conclude by Lemma 1.8 that $\{U_{t,n}\}$ is strongly convergent if $t \ge 0$. (For $t < 0$ the proof is essentially the same.) We define

$$U_t := \operatorname*{s-lim}_{n\to\infty} U_{t,n} \quad (4.23)$$

and notice that

$$\|U_t\| \le 1. \quad (4.24)$$

(iii) Let $g \in D(A^2)$. We may write

$$\|U_t g - U_s g\| \le \|U_t g - U_{t,n} g\| + \|U_{t,n} g - U_{s,n} g\| + \|U_{s,n} g - U_s g\| .$$

Let $\varepsilon > 0$. From (4.22) we see that one may choose n such that the first and the third term on the r.h.s. are each $<\varepsilon/3$, for all s satisfying $|s| \le |t| + 1$. Since $t \mapsto U_{t,n}$ is strongly continuous, the second term is $<\varepsilon/3$ if $|s - t| < \delta$ for some $\delta > 0$. This shows that U_t is *strongly continuous* (using again Lemma 1.8).

(iv) Since A is self-adjoint, we have (see also (4.48))

$$U^*_{t,n} = U_{-t,n}. \quad (4.25)$$

Also, for $f, g \in D(A)$ (cf. Proposition 2.9 (a))

$$(Ag, U_t f) = \lim_{n\to\infty} (Ag, U_{t,n} f) = \lim_{n\to\infty} (U_{-t,n} Ag, f)$$

$$= \lim_{n\to\infty} (AU_{-t,n} g, f) = \lim_{n\to\infty} (g, U_{t,n} Af) = (g, U_t Af).$$

Hence $U_t f \in D(A^*) = D(A)$ and

$$A^* U_t f = AU_t f = U_t Af. \quad (4.26)$$

In particular U_t leaves D(A) invariant.

(v) For f∈D(A), we have from (4.20) and (4.21) that

$$\frac{d}{dt}U_{t,n}f = -iU_{t,n}(\frac{it}{n}A + I)^{-1}Af$$

and

$$\underset{n\to\infty}{\text{s-lim}}\ \frac{d}{dt}U_{t,n}f = -iU_tAf. \tag{4.27}$$

Since $||\frac{d}{dt}U_{t,n}f + iU_tAf|| \leq 2||Af||$, we obtain from the Lebesgue dominated convergence theorem that

$$||\int_0^s(\frac{d}{dt}U_{t,n}f + iU_tAf)dt|| \leq \int_0^s||\frac{d}{dt}U_{t,n}f + iU_tAf||dt \to 0$$

as $n \to \infty$. Thus

$$-i\int_0^s U_tAfdt = \underset{n\to\infty}{\text{s-lim}} \int_0^s \frac{d}{dt}U_{t,n}f\,dt = \underset{n\to\infty}{\text{s-lim}}(U_{s,n}f - f) = U_sf - f.$$

Since the integrand is strongly continuous, this means that

$$\frac{d}{ds}U_sf = -iU_sAf \qquad \forall\, f\in D(A). \tag{4.28}$$

(vi) Let f∈D(A). Then by (4.28) and (4.26),

$$\frac{d}{ds}(g,U_{t-s}U_sf) = i(g,U_{t-s}AU_sf) - i(g,U_{t-s}U_sAf) = 0.$$

It follows that $s \mapsto (g,U_{t-s}U_sf)$ is constant for each f∈D(A) and fixed t. Therefore, for fixed t, $U_{t-s}U_sf$ is a constant vector. Since this holds for each f∈D(A), we have $U_{t-s}U_s = B_t$, where B_t is an operator in $\mathcal{B}(\mathcal{H})$ independent of s. Setting $s = 0$, we find that $B_t = U_t$. Hence $U_{t-s}U_s = U_t$. Setting $t = \tau + s$, one obtains that

$$U_\tau U_s = U_{\tau+s}, \tag{4.29}$$

i.e. $\{U_t\}$ is a *one-parameter group*.

(vii) Setting $\tau = -s$ in (4.29), one may deduce that

$$U_t U_{-t} = I = U_{-t} U_t \quad \forall t \in \mathbb{R}. \tag{4.30}$$

We show that $||U_t f|| = ||f||$ for each $f \in H$. Assume to the contrary that $||U_t f|| < ||f||$ for some f. Then $||f|| = ||U_{-t}U_t f|| \leq ||U_t f|| < ||f||$, a contradiction.

Now, using the polarization identity (1.14) and the fact that $||U_t f|| = ||f||$, one may deduce that $(g, U_t^* U_t f) = (U_t g, U_t f) = (g, f)$ for all $f, g \in H$. Thus $U_t^* U_t f - f$ is orthogonal to H, hence $U_t^* U_t f = f$. This means that U_t is isometric. Since $R(U_t) = H$ by (4.30), it follows that U_t is *unitary* (cf. Proposition 1.17).

(viii) Denote by A_o the infinitesimal generator of $\{U_t\}$. By (4.28), A_o is a (self-adjoint) extension of A; since A is self-adjoint, one must have $A_o = A$.

The uniqueness of the group $\{U_t\}$ is easily obtained. Assume $\{\tilde{U}_t\}$ is another group having A as its infinitesimal generator. Then one deduces as in (vi) that $U_{t-s}\tilde{U}_s = C_t$ for some $C_t \in B(H)$. Taking $s = 0$ and $s = t$, one obtains $U_t = C_t = \tilde{U}_t$. □

Corollary 4.5 : If f is an eigenvector of A, $Af = \lambda f$, then $U_t f = \exp(-i\lambda t) f$.

Proof : Clearly $(\alpha A + I)^{-1} f = (\alpha\lambda + 1)^{-1} f$. Hence $U_{t,n} f = (it\lambda n^{-1} + 1)^{-n} f$, which converges as $n \to \infty$ to $\exp(-i\lambda t) f$. □

We saw in Section 2.2 that a self-adjoint operator A induces an orthogonal decomposition of the Hilbert space H into two subspaces $H_p(A)$ and $H_c(A)$. In the next proposition we establish some relations between this decomposition of H and the evolution group associated with A.

Proposition 4.6 : Let $\{U_t\}$ be an evolution group, A its infinitesimal generator and $H = H_p(A) \oplus H_c(A)$ the associated decomposition of

$\mathcal{H}$. Denote by $E_p(A)$ and $E_c(A)$ the orthogonal projections with range $\mathcal{H}_p(A)$ and $\mathcal{H}_c(A)$ respectively. Then

(a) If $B \in \mathcal{B}(\mathcal{H})$ and $BU_t = U_t B$ for all $t \in \mathbb{R}$, then the restrictions of B to $\mathcal{H}_p(A)$ and $\mathcal{H}_c(A)$ leave $\mathcal{H}_p(A)$ and $\mathcal{H}_c(A)$ respectively invariant, i.e. $B = B_p \oplus B_c$, where B_p is an operator acting in $\mathcal{H}_p(A)$ and B_c an operator acting in $\mathcal{H}_c(A)$. In other words, if $f = f_p \oplus f_c$, then

$$Bf = B_p f_p \oplus B_c f_c \ , \tag{4.31}$$

or equivalently

$$BE_p(A) = E_p(A)B, \qquad BE_c(A) = E_c(A)B. \tag{4.32}$$

(b) U_t leaves $\mathcal{H}_p(A)$ and $\mathcal{H}_c(A)$ invariant, and the restrictions of U_t to $\mathcal{H}_p(A)$ and $\mathcal{H}_c(A)$ define strongly continuous unitary one-parameter groups $U_{t,p}$ and $U_{t,c}$ acting in $\mathcal{H}_p(A)$ and $\mathcal{H}_c(A)$ respectively.

Proof : (a) The hypothesis $BU_t = U_t B$ implies that $U_t^* B^* = B^* U_t^*$ for each $t \in \mathbb{R}$. This implies by (4.5) that $U_s B^* = B^* U_s$ for each $s \in \mathbb{R}$. Hence B^* also commutes with the group $\{U_t\}$.

Let f be an eigenvector of A : $Af = \lambda f$. Since $BU_t f = U_t Bf$ and $BU_t f$ is strongly differentiable, we have that $Bf \in D(A)$ and $-iB\lambda f = -iABf$. In other words, $A(Bf) = \lambda(Bf)$. Thus $Bf \in \mathcal{H}_p(A)$ for each eigenvector f of A. By the linearity of B and the definition of $\mathcal{H}_p(A)$, this implies that $Bf \in \mathcal{H}_p(A)$ for each $f \in \mathcal{H}_p(A)$, which means that B leaves $\mathcal{H}_p(A)$ invariant. By the observation made at the beginning, the same is true for B^* : $B^* f \in \mathcal{H}_p(A)$ for each $f \in \mathcal{H}_p(A)$.

Now let $g \in \mathcal{H}_c(A)$. Then, for each $f \in \mathcal{H}_p(A)$, we have $(Bg,f) = (g,B^*f) = 0$, since $B^*f \in \mathcal{H}_p(A)$. Hence $Bg \perp \mathcal{H}_p(A)$, i.e. $Bg \in \mathcal{H}_c(A)$. Hence B also leaves $\mathcal{H}_c(A)$ invariant, and we get (4.31).

Since we also have $B^* = B_p^* \oplus B_c^*$, (4.31) implies that

$$(B^*B)_p = B^*_p B_p \ , \qquad (B^*B)_c = B^*_c B_c . \tag{4.33}$$

(b) By (4.3), U_s commutes with U_t for all t. Hence, by (a), $U_s = U_{s,p} \oplus U_{s,c}$ and $U^*_s = U^*_{s,p} \oplus U^*_{s,c}$. The fact that $\{U_{t,p}\}$ and $\{U_{t,c}\}$ are unitary one-parameter groups follows from (4.33). For example :

$$U^*_{t,p} U_{t,p} = (U^*_t U_t)_p = I_p = E_p(A) = (U_t U^*_t)_p = U_{t,p} U^*_{t,p}$$

and $U_{t,p} U_{s,p} = (U_t U_s)_p = U_{t+s,p}$. □

The next lemma gives another result on commutativity.

Lemma 4.7 : Let $\{U_t\}$ be an evolution group, A its infinitesimal generator. Assume that $B \in \mathcal{B}(\mathcal{H})$ is self-adjoint and that $BU_t = U_t B$ for all $t \in \mathbb{R}$. Then

(a) U_t leaves each eigensubspace $\mathcal{M}_i$ of B and $\mathcal{H}_c(B)$ invariant, i.e., in the decomposition $\mathcal{H} = \bigoplus_i \mathcal{M}_i \oplus \mathcal{H}_c(B)$:

$$U_t = \bigoplus_i U_{t;i} \oplus U_{t;c} . \tag{4.34}$$

(b) The operators $\{U_{t;i}\}$ form a strongly continuous unitary one-parameter group in $\mathcal{M}_i$.

(c) $D(A) \cap \mathcal{M}_i$ is dense in $\mathcal{M}_i$.

Proof : (i) Let $f \in \mathcal{M}_i$: $Bf = \lambda_i f$. Then $BU_t f = U_t Bf = \lambda_i U_t f$, so that $U_t f \in \mathcal{M}_i$.

If $g \in \mathcal{H}_c(B)$, we have $(U_t g, f) = (g, U_{-t} f) = 0$ for each $f \in \mathcal{H}_p(B)$, since $U_{-t} f \in \mathcal{H}_p(B)$. Thus U_t leaves $\mathcal{H}_c(B)$ invariant and we have (4.34). The proof that $\{U_{t;i}\}$ is a unitary one-parameter group is as in (b) of the preceding proof.

(ii) The infinitesimal generator of $\{U_{t;i}\}$ is a densely defined self-adjoint operator A_i in $\mathcal{M}_i$. By (4.34), $A_i f = Af$ for each

$f \in D(A_i)$, hence $D(A_i) \subseteq D(A) \cap M_i$. This proves the assertion of (c). □

Remark : The lemma is of course also true if B is unbounded, but in that case the hypothesis of commutativity of B and U_t is not so easy to formulate.

Proposition 4.8 : Let $\{U_t\}$ be an evolution group and A its infinitesimal generator. For $f \in \mathcal{H}$, define its orbit $\mathcal{O}(f)$ by $\mathcal{O}(f) :=$ $= \{U_s f \mid s \in \mathbb{R}\}$, and denote by $\mathcal{M}(f)$ the subspace spanned by $\mathcal{O}(f)$. Then, if $f \in \mathcal{H}_c(A)$ and $f \neq \theta$, $\mathcal{M}(f)$ is infinite-dimensional.

Proof : Clearly each U_t leaves $\mathcal{O}(f)$ invariant. By Lemma 1.18, U_t leaves $\mathcal{M}(f)$ invariant. By taking $B = E(f)$ (the projection with range $\mathcal{M}(f)$) in Lemma 4.7, we see that the restriction of U_t to $\mathcal{M}(f)$ defines an evolution group in $\mathcal{M}(f)$. ($\mathcal{M}(f)$ is the eigensubspace corresponding to the eigenvalue $\lambda = 1$ of $E(f)$.) Its infinitesimal generator coincides with the restriction of A to $\mathcal{M}(f)$. If $\mathcal{M}(f)$ were finite-dimensional, the restriction of A to $\mathcal{M}(f)$ would have pure point spectrum (in a finite-dimensional Hilbert space, every self-adjoint operator has pure point spectrum). Since $\mathcal{M}(f) \subseteq \mathcal{H}_c(A)$, this is impossible, whence $\dim \mathcal{M}(f) = \infty$. □

4.2 Functional Calculus

Let $\{U_t\}$ be an evolution group and A its infinitesimal generator. The purpose of this section is to define functions of A and some of their properties.

The standard way of defining functions of self-adjoint operators is to use the spectral theorem. We want to avoid the spectral theorem here and therefore use a definition directly in terms of the evolution group. This definition gives only a restricted class of functions of A, which is however sufficient for the applications we have in mind.

Suppose $\varphi:\mathbb{R}\to\mathbb{C}$ is the inverse Fourier transform of a function $\tilde{\varphi}$ in $L^1(\mathbb{R})$ (hence, formally, $\tilde{\varphi}$ is the Fourier transform of φ). Consider the expression

$$(2\pi)^{-1/2}\int_{-\infty}^{\infty}\tilde{\varphi}(t)U_t^*dt. \tag{4.35}$$

By Proposition 1.24, this defines an operator in $\mathcal{B}(H)$, since

$$\int_{-\infty}^{\infty}\|\tilde{\varphi}(t)U_t^*\|\,dt = \int_{-\infty}^{\infty}|\tilde{\varphi}(t)|\,dt = \|\tilde{\varphi}\|_1 < \infty. \tag{4.36}$$

If $\tilde{\varphi}$ is continuous, the integral in (4.35) is defined as a Riemann integral, as explained in Section 1.3. If $\tilde{\varphi}$ is not continuous, one may use a more general definition of the integral [HP]; in later applications $\tilde{\varphi}$ will always be continuous.

The operator defined by (4.35) will be denoted $\varphi(A)$:

$$\varphi(A) := (2\pi)^{-1/2}\int_{-\infty}^{\infty}\tilde{\varphi}(t)U_t^*dt. \tag{4.37}$$

This may be justified formally by writing $U_t^* = \exp(iAt)$, so that, again formally, (4.37) gives the inverse Fourier transform φ of $\tilde{\varphi}$, the argument being the operator A. Of course the definition (4.37) of $\varphi(A)$ gives the same operator as the definition in terms of the spectral family of A, which, as already said, we do not use in these lectures.

We shall now prove some simple properties of the operators $\varphi(A)$.

<u>Proposition 4.9</u> : Assume $\tilde{\varphi},\tilde{\psi}\in L^1(\mathbb{R})$. Then

(a) $\varphi(A)\in\mathcal{B}(H)$ and $\|\varphi(A)\| \le (2\pi)^{-1/2}\|\tilde{\varphi}\|_1$.

(b) $\varphi(A)U_s = U_s\varphi(A) \qquad \forall\, s\in\mathbb{R}.$ (4.38)

(c) $[\varphi(A)]^* = \bar{\varphi}(A)$, where $\bar{\varphi}(\lambda) := \overline{\varphi(\lambda)}$. (4.39)

(d) $\varphi(A)\psi(A) = \psi(A)\varphi(A) = (\varphi\psi)(A)$, (4.40)

where $(\varphi\psi)(\lambda) := \varphi(\lambda)\psi(\lambda)$.

(e) If $Af = \lambda f$, then $\varphi(A)f = \varphi(\lambda)f$.

(f) If $f\in D(A)$, then $\varphi(A)f\in D(A)$ and

$$A\varphi(A)f = \varphi(A)Af. \tag{4.41}$$

(g) $f\in H_p(A) \Rightarrow \varphi(A)f\in H_p(A)$, and $g\in H_c(A) \Rightarrow \varphi(A)g\in H_c(A)$, or equivalently

$$E_p(A)\varphi(A) = \varphi(A)E_p(A) \quad , \quad E_c(A)\varphi(A) = \varphi(A)E_c(A). \tag{4.42}$$

<u>Proof</u> : (a) follows from Proposition 1.24 and (4.36), (e) from the definition (4.37) and Corollary 4.5, and (b) is immediate.

(c) Since $\overline{\tilde{\varphi}(t)} = \tilde{\bar{\varphi}}(-t)$, one gets by using (4.5) and setting $\tau = -t$:

$$(2\pi)^{1/2}\varphi(A)^* = \int_{-\infty}^{\infty}\overline{\tilde{\varphi}(t)}U_t dt = \int_{-\infty}^{\infty}\tilde{\bar{\varphi}}(-t)U_t dt$$

$$= \int_{-\infty}^{\infty}\tilde{\bar{\varphi}}(\tau)U_\tau^* d\tau = (2\pi)^{1/2}\bar{\varphi}(A).$$

(d) One has

$$(\widetilde{\varphi\psi})(t) = (\tilde{\varphi}*\tilde{\psi})(t) = (2\pi)^{-1/2}\int\tilde{\varphi}(t-s)\tilde{\psi}(s)ds.$$

This shows that $\widetilde{\varphi\psi}\in L^1(\mathbb{R})$, so that $(\varphi\psi)(A)$ is well defined. Now, for all $f,g\in H$

$$(f,\varphi(A)\psi(A)g) - (f,(\varphi\psi)(A)g) = (2\pi)^{-1}\iint(f,U_t^*U_s^*g)\tilde{\varphi}(t)\tilde{\psi}(s)dtds$$

$$- (2\pi)^{-1}\iint(f,U_t^*g)\tilde{\varphi}(t-s)\tilde{\psi}(s)dtds,$$

where the order of integration is immaterial by Fubini's theorem. By making the change of variables $t \mapsto \tau - s$ in the first integral, one obtains that $(f,\varphi(A)\psi(A)g) = (f,(\varphi\psi)(A)g)$ for all $f,g\in H$, and the result of (d) now follows in the usual way from Lemma 1.5.

(f) If $f\in D(A)$, then

$$\frac{1}{s}\Big[U_s - I\Big]\varphi(A)f + i\varphi(A)Af = \varphi(A)[s^{-1}(U_s - I) + iA]f,$$

which converges strongly to zero as $s \to 0$. Hence, by Proposition 4.1, $\varphi(A)f\in D(A)$ and $A\varphi(A)f = \varphi(A)Af$.

(g) By Proposition 4.6, U_t^* leaves $H_p(A)$ and $H_c(A)$ invariant, and the assertions of (g) follow from this and the definition (4.37).□

<u>Lemma 4.10</u> : Assume that $\tilde{\varphi}\in L^1(\mathbb{R})$. Then

$$\underset{\tau\to 0}{\text{s-lim}}\int_{\mathbb{R}}\tilde{\varphi}(s)U_{\tau s}^* ds = (2\pi)^{1/2}\varphi(0)I. \tag{4.43}$$

<u>Proof</u> : One has $(2\pi)^{1/2}\varphi(0) = \int_{-\infty}^{\infty}\tilde{\varphi}(s)ds$. Now, for each $f\in H$,

$$\left\|\int_{-\infty}^{\infty}\tilde{\varphi}(s)(U_{\tau s}^* - I)f ds\right\| \le \int_{-\infty}^{\infty}|\tilde{\varphi}(s)|\,\|(U_{\tau s}^* - I)f\|ds.$$

For each fixed s with $|\tilde{\varphi}(s)| < \infty$, the integrand converges to zero as $\tau \to 0$, by the strong continuity of $\{U_t^*\}$. Furthermore it is majorized, uniformly in $\tau\in\mathbb{R}$, by the L^1-function $2|\tilde{\varphi}(s)|\,\|f\|$. Hence, by the Lebesgue dominated convergence theorem, the integral on the r.h.s. converges to zero as $\tau \to 0$, which proves (4.43). □

<u>Proposition 4.11</u> : Assume that $\tilde{\varphi}\in L^1(\mathbb{R})$, $\tilde{\varphi}$ is continuously differentiable and that its derivative $\tilde{\varphi}'$ also belongs to $L^1(\mathbb{R})$. Then $R(\varphi(A))\subseteq D(A)$, $A\varphi(A)$ is in $B(H)$ and given by

$$A\varphi(A) = i(2\pi)^{-1/2}\int_{-\infty}^{\infty}\tilde{\varphi}'(t)U_t^* dt. \tag{4.44}$$

<u>Proof</u> : (i) Since $\tilde{\varphi}\in L^1(\mathbb{R})$, there exist two sequences $\{r_i^{\pm}\}$ of real numbers such that $r_i^+ \to +\infty$, $r_i^- \to -\infty$ and $\tilde{\varphi}(r_i^{\pm}) \to 0$ as $i \to \infty$ (otherwise clearly $|\tilde{\varphi}|$ could not be integrable). By using Proposition

1.22 (c), we then obtain that, for $f \in D(A)$:

$$\int_{r_i^-}^{r_i^+} \frac{d}{dt}\left[\tilde{\varphi}(t)U_t^* f\right]dt = \tilde{\varphi}(r_i^+)U_{r_i^+}^* f - \tilde{\varphi}(r_i^-)U_{r_i^-}^* f \to \theta \quad \text{as} \quad i \to \infty.$$

By calculating the derivative on the l.h.s., we get from this that

$$\int_{-\infty}^{\infty}\tilde{\varphi}'(t)U_t^* f\,dt + i\int_{-\infty}^{\infty}\tilde{\varphi}(t)U_t^* Af\,dt = \theta, \tag{4.45}$$

where both integrals exist in $\mathcal{B}(\mathcal{H})$ by our assumptions on $\tilde{\varphi}$. (4.45) and (4.41) imply that (4.44) holds on $D(A)$.

(ii) Let $f \in \mathcal{H}$. Choose a sequence $\{f_n\} \in D(A)$ such that $\text{s-lim}\, f_n = f$. Since $\varphi(A)$ is in $\mathcal{B}(\mathcal{H})$, $\text{s-lim}\, \varphi(A)f_n = \varphi(A)f$. By (i), we have $\text{s-lim}\, A\varphi(A)f_n = \text{s-lim}\, i\psi(A)f_n = i\psi(A)f$, where

$$\psi(A) := (2\pi)^{-1/2}\int_{-\infty}^{\infty}\tilde{\varphi}'(t)U_t^* dt \in \mathcal{B}(\mathcal{H}).$$

Since A is closed, this implies that $\varphi(A)f \in D(A)$ and $A\varphi(A)f = i\psi(A)f$. Thus $A\varphi(A)$ is everywhere defined, bounded and satisfies (4.44). □

Examples of functions satisfying all hypotheses of Propositions 4.9 and 4.11 are functions φ in $C_0^\infty(\mathbb{R})$. The resolvent $(A - z)^{-1}$ is also of the form $\varphi(A) \equiv \psi_z(A)$; in fact, by Corollary 4.2, we have :

$$\text{If } \operatorname{Im} z > 0 : \tilde{\psi}_z(t) = \sqrt{2\pi}\, i e^{-izt}\chi_{(-\infty,0)}(t), \tag{4.46}$$

$$\text{if } \operatorname{Im} z < 0 : \tilde{\psi}_z(t) = -\sqrt{2\pi}\, i e^{-izt}\chi_{(0,\infty)}(t), \tag{4.47}$$

where χ_Δ denotes the characteristic function of the set Δ. Notice that $\tilde{\psi}_z \in L^1(\mathbb{R})$ but that its derivative is not a function. Also, (4.46) and (4.47) imply together with (4.39) that

$$[(A - z)^{-1}]^* = (A - \bar{z})^{-1}. \tag{4.48}$$

Given a function φ, we may define a sequence $\{\varphi_\varepsilon\}_{\varepsilon>0}$ by adding an exponential convergence factor in the Fourier transform $\tilde{\varphi}$ of φ. More precisely, we define φ_ε by

$$(\tilde{\varphi_\varepsilon})(t) = \tilde{\varphi}(t)e^{-\varepsilon|t|} \qquad (\varepsilon > 0). \tag{4.49}$$

In the next lemma we relate $\varphi_\varepsilon(A)$ to the resolvent of A and show that $\varphi_\varepsilon(A)$ converges to $\varphi(A)$ as $\varepsilon \to 0$.

Lemma 4.12 : Assume that φ and $\tilde{\varphi}$ are in $L^1(\mathbb{R})$. Then

$$\varphi_\varepsilon(A) = (2\pi i)^{-1}\int_{-\infty}^{\infty}\varphi(\lambda)[(A-\lambda-i\varepsilon)^{-1} - (A-\lambda+i\varepsilon)^{-1}]d\lambda, \tag{4.50}$$

and

$$\lim_{\varepsilon\to 0} \|\varphi(A) - \varphi_\varepsilon(A)\| = 0. \tag{4.51}$$

Proof : (i) We have $\|(A-\lambda \pm i\varepsilon)^{-1}\| \leq \varepsilon^{-1}$ by (2.14), and $\lambda \mapsto (A - \lambda \pm i\varepsilon)^{-1}$ is continuous in the operator norm by (2.17). Since $\varphi\in L^1(\mathbb{R})$, this shows together with Proposition 1.24 (d) that the integral in (4.50) exists.

For any $f,g\in H$ we have by Corollary 4.2 :

$$\int_{-\infty}^{\infty}d\lambda\varphi(\lambda)(f,[(A-\lambda-i\varepsilon)^{-1} - (A-\lambda+i\varepsilon)^{-1}]g)$$

$$= i\int_{-\infty}^{\infty}d\lambda\varphi(\lambda)\int_{-\infty}^{\infty}dt e^{i\lambda t}e^{-\varepsilon|t|}(f,U_t g).$$

The integrand in this double integral is majorized by the integrable function $|\varphi(\lambda)|\exp(-\varepsilon|t|)\|f\|\|g\|$. Hence we may interchange the order of integration. Making also the change of variables $t \mapsto -t$, we find that the last expression is equal to

$$i\sqrt{2\pi}\int_{-\infty}^{\infty}\tilde{\varphi}(t)e^{-\varepsilon|t|}(f,U_t^* g)dt = 2\pi i(f,\varphi_\varepsilon(A)g).$$

Since this holds for all $f,g\in H$, we have (4.50) (use Lemma 1.5).

(ii) $\|\varphi(A) - \varphi_\varepsilon(A)\| \le (2\pi)^{-1/2}\|\tilde{\varphi} - \tilde{\varphi}_\varepsilon\|_1$

$$= (2\pi)^{-1/2}\int_{-\infty}^{\infty}|\tilde{\varphi}(t)|(1 - e^{-\varepsilon|t|})dt$$

by Proposition 4.9 (a). Since $\tilde{\varphi}\in L^1(\mathbb{R})$, this integral converges to zero as $\varepsilon \to 0$ by the Lebesgue dominated convergence theorem. This proves (4.51). □

The remainder of this section is devoted to proving part (a) of Proposition 3.12, namely that a compact self-adjoint operator cannot have continuous spectrum. We begin with a few preliminary results.

We pick a function $\varphi\in C_0^\infty(\mathbb{R})$ having the following properties :

(i) $|\varphi(\lambda)| \le 1 \quad \forall\,\lambda\in\mathbb{R}$,

(ii) $\varphi(\lambda) = 1$ if $|\lambda| \le 1$, (4.52)

(iii) $\varphi(\lambda) = 0$ if $|\lambda| \ge 2$.

For given $\mu\in\mathbb{R}$, we define a sequence $\{\varphi_{\mu,n}\}$ of functions in $C_0^\infty(\mathbb{R})$ by

$$\varphi_{\mu,n}(\lambda) = \varphi(2^n(\lambda - \mu)). \tag{4.53}$$

It is easily checked that these functions have the following properties :

$$\varphi_{\mu,n}(\lambda)\varphi_{\mu,n+m}(\lambda) = \varphi_{\mu,n+m}(\lambda) \qquad \forall\, m \ge 1, \tag{4.54}$$

$$\tilde{\varphi}_{\mu,n}(t) = 2^{-n}e^{-i\mu t}\tilde{\varphi}(2^{-n}t), \tag{4.55}$$

$$\frac{d}{dt}\tilde{\varphi}_{\mu,n}(t) := \tilde{\varphi}'_{\mu,n}(t) = -i\mu\tilde{\varphi}_{\mu,n}(t) + 2^{-2n}e^{-i\mu t}\tilde{\varphi}'(2^{-n}t). \tag{4.56}$$

Lemma 4.13 : Let A be self-adjoint. Then, for each $\mu\in\mathbb{R}$,

$$\lim_{n\to\infty}\|(A - \mu)\varphi_{\mu,n}(A)\| = 0. \tag{4.57}$$

Proof : By (4.44) and (4.56) we have

$$(2\pi)^{1/2}(A - \mu)\varphi_{\mu,n}(A) = i\int_{-\infty}^{\infty}\tilde{\varphi}'_{\mu,n}(t)U_t^*dt - \mu\int_{-\infty}^{\infty}\tilde{\varphi}_{\mu,n}(t)U_t^*dt$$

$$= 2^{-2n}i\int_{-\infty}^{\infty}e^{-i\mu t}\tilde{\varphi}'(2^{-n}t)U_t^*dt = 2^{-n}i\int_{-\infty}^{\infty}\exp(-i2^n\mu\tau)\tilde{\varphi}'(\tau)U^*_{2^n\tau}d\tau,$$

where we have made the change of variables $\tau = 2^{-n}t$. It follows that

$$(2\pi)^{1/2}\|(A - \mu)\varphi_{\mu,n}(A)\| \leq 2^{-n}\int_{-\infty}^{\infty}|\tilde{\varphi}'(\tau)|d\tau,$$

which converges to zero as $n \to \infty$. □

Remark 4.14 : (a) If μ is in the resolvent set of A, then we may deduce that $\|\varphi_{\mu,n}(A)\| \to 0$ as $n \to \infty$. In fact :

$$\|\varphi_{\mu,n}(A)\| \leq \|(A - \mu)^{-1}\| \, \|(A - \mu)\varphi_{\mu,n}(A)\| \to 0$$

by Lemma 4.13. (This result is of course not optimal, since one knows that $\varphi_{\mu,n}(A) = 0$ for all $n \geq n_o(\mu)$ if $\mu \in \rho(A)$.)

(b) If $f \in \mathcal{H}_p(A)$, then s-lim $\varphi_{\mu,n}(A)f = E_\mu f$, where E_μ is the orthogonal projection whose range is the eigensubspace associated with μ (which is non-zero only if μ is an eigenvalue of A). The proof is easily obtained by using Proposition 4.9 (e).

Lemma 4.15 : Let A be self-adjoint, and let $\mathcal{R}_{\mu,n}$ be the range of $\varphi_{\mu,n}(A)$. Then

(a) $U_t\mathcal{R}_{\mu,n} \subseteq \mathcal{R}_{\mu,n} \qquad \forall\, t \in \mathbb{R},$ (4.58)

(b) $\mathcal{R}_{\mu,n+m} \subseteq \mathcal{R}_{\mu,n} \qquad \forall\, m \geq 0.$ (4.59)

Proof : (a) Follows immediately from (4.38) and (b) from (4.54) and (4.40). □

Lemma 4.16 : Let A be self-adjoint and have purely continuous spectrum. Define $M_{\mu,n}$ to be the closure of $R_{\mu,n}$. Then the subspace $M_{\mu,n}$ is either infinite-dimensional or zero.

Proof : If $R_{\mu,n} \neq \{\theta\}$, there is a vector $f \in H$ such that $\varphi_{\mu,n}(A)f \neq \theta$. Then, by (4.58), $R_{\mu,n}$ contains the orbit $\{U_t \varphi_{\mu,n}(A)f\}$, which spans an infinite-dimensional subspace by Proposition 4.8. □

Lemma 4.17 : Let A be self-adjoint and have purely continuous spectrum, and let $\mu \in \mathbb{R}$. Then, either there is an integer n_o such that $\varphi_{\mu,n}(A) = 0$ for all $n \geq n_o$, or there is a sequence of vectors $\{g_n\} \in D(A)$ satisfying $||g_n|| = 1$, w-lim $g_n = \theta$ and s-lim$(A - \mu)g_n = \theta$.

Proof : Since $R_{\mu,n+1} \subseteq R_{\mu,n}$, Lemma 4.16 implies that either there is a number n_o such that $R_{\mu,n} = \{\theta\}$ for all $n \geq n_o$, or dim $R_{\mu,n} = \infty$ for all n. In the first case, $\varphi_{\mu,n}(A) = 0$ for all $n \geq n_o$. In the second case, we shall construct $\{g_n\}$ and two auxiliary sequences $\{f_n\}$ and $\{h_n\}$ by recursion. $\{h_n\}$ will be an infinite orthonormal sequence.

We take f_1 such that $\varphi_{\mu,1}(A)f_1 \neq \theta$ and set $g_1 = h_1 = $
$= \varphi_{\mu,1}(A)f_1/||\varphi_{\mu,1}(A)f_1||$. The recursion is as follows : given $f_1,\ldots,f_n$ and $h_1,\ldots,h_n$, we choose $h_{n+1} \in \overline{R_{\mu,n+1}}$ such that $||h_{n+1}|| = 1$ and such that h_{n+1} is orthogonal to $h_1,\ldots,h_n$, which is possible by Lemma 1.4. Since $R_{\mu,n+1}$ is a linear manifold, hence dense in its closure, we can find a vector f_{n+1} in H such that

$$||h_{n+1} - \varphi_{\mu,n+1}(A)f_{n+1}|| < 1/(n + 1). \qquad (4.60)$$

Finally we set $g_{n+1} = \varphi_{\mu,n+1}(A)f_{n+1}/||\varphi_{\mu,n+1}(A)f_{n+1}||$.

Clearly $||g_n|| = 1$ for each n and $g_n \in D(A)$. Furthermore, (4.60) implies that $1 - 1/n < ||\varphi_{\mu,n}(A)f_n|| < 1 + 1/n$ (use the inequality (1.15)). Hence

$$\lim_{n\to\infty} \|\varphi_{\mu,n}(A)f_n\| = 1. \qquad (4.61)$$

Thus, for each $f \in H$:

$$(f,g_n) = (f,h_n) - (f,[h_n - \varphi_{\mu,n}(A)f_n])$$
$$- (f,[1 - \|\varphi_{\mu,n}(A)f_n\|^{-1}]\varphi_{\mu,n}(A)f_n).$$

The first term on the r.h.s. converges to zero by Example 1.2, the second term by (4.60) (apply the Schwarz inequality) and the third one by (4.61). Hence w-lim $g_n = \theta$.

Finally, by (4.54) :

$$g_n = \varphi_{\mu,n-1}(A)\varphi_{\mu,n}(A)f_n/\|\varphi_{\mu,n}(A)f_n\| = \varphi_{\mu,n-1}(A)g_n.$$

Hence $\|(A-\mu)g_n\| \le \|(A-\mu)\varphi_{\mu,n-1}(A)\|\,\|g_n\|$, which converges to zero by Lemma 4.13, since $\|g_n\| = 1$. □

Proof of Proposition 3.12 (a) : Let A be self-adjoint and compact, and consider the Hilbert space $K = H_c(A)$. All operators below are assumed to act in K.

(i) We first show that, for each $\mu \neq 0$, there is an integer $n_0(\mu) < \infty$ such that $\varphi_{\mu,n}(A) = 0$ for all $n \ge n_0(\mu)$. If this were not the case for some $\mu \neq 0$, there would be a sequence $\{g_n\}$ in K such that $\|g_n\| = 1$, w-lim $g_n = \theta$ and s-lim$(A-\mu)g_n = \theta$, by Lemma 4.17. Since A is compact, we would then have s-lim $Ag_n = \theta$ by Proposition 3.9, hence s-lim $\mu g_n = \theta$. Since $\mu \neq 0$ and $\|g_n\| = 1$, this is impossible, which proves our claim.

(ii) Let $\mu \neq 0$ and $\Delta(\mu) = \{\lambda \in \mathbb{R} \,|\, |\lambda-\mu| < 2^{-n_0(\mu)}\}$. If $\psi \in C_0^\infty(\mathbb{R})$ is such that it vanishes outside $\Delta(\mu)$, we have by (4.40) and (4.53) that

$$\psi(A) = \psi(A)\varphi_{\mu,n_0(\mu)}(A) = 0. \qquad (4.62)$$

(iii) Let φ be the function introduced in (4.52), and set $\chi_m(\lambda) = \varphi(\lambda/m)$. Then $\tilde{\chi}_m(t) = m\tilde{\varphi}(mt)$ and

$$\chi_m(A) = (2\pi)^{-1/2} m\int_{-\infty}^{\infty}\tilde{\varphi}(mt)U_t^* dt = (2\pi)^{-1/2}\int_{-\infty}^{\infty}\tilde{\varphi}(s)U_{s/m}^* ds.$$

Hence, by Lemma 4.10,

$$\underset{m\to\infty}{\text{s-lim}}\ \chi_m(A) = I. \tag{4.63}$$

(iv) Let $f\in K$ and $\varepsilon > 0$. Choose m such that

$$\|\chi_m(A)f - f\| < \varepsilon. \tag{4.64}$$

The support of $\chi_m - \varphi_{o,n}$ is $S_{mn} = \{\lambda\in\mathbb{R} \mid 2^{-n} \le |\lambda| \le 2m\}$. Each point μ in S_{mn} contains an open neighbourhood $\Delta(\mu)$ such that $\psi(A) = 0$ if $\text{supp}\ \psi\subset\Delta(\mu)$. Choose a finite set of points $\mu_i\in S_{mn}$ such that the corresponding intervals $\Delta(\mu_i)$ form an open covering of S_{mn}, which is possible by the compactness of S_{mn}. Then one may write $\chi_m - \varphi_{o,n} = \sum_i\psi_{mn}^{(i)}$, where $\psi_{mn}^{(i)}\in C_o^\infty(\mathbb{R})$ and $\text{supp}\ \psi_{mn}^{(i)}\subset\Delta(\mu_i)$. It follows that $\chi_m(A) - \varphi_{o,n}(A) = \sum_i\psi_{mn}^{(i)}(A) = 0$. Hence, using (4.64) : $\|\varphi_{o,n}(A)f - f\| < \varepsilon$ for each n and each $\varepsilon > 0$. Consequently $\varphi_{o,n}(A)f = f$ for each n, and hence s-lim $\varphi_{o,n}(A)f = f$ as $n \to \infty$. This in turn implies that

$$Af = \underset{n\to\infty}{\text{s-lim}}\ A\varphi_{o,n}(A)f = \theta,$$

by Lemma 4.13. Thus A is the zero operator in K. Since the zero operator has pure point spectrum and we had $K = H_c(A)$, we have a contradiction unless $K = H_c(A) = \{\theta\}$. □

4.3 Ergodic Properties of Evolution Groups

In this section we give a few results about ergodic properties of evolution groups. These will be very useful in Chapter 5 for studying the asymptotic behaviour of the part of an evolution group in the subspace of continuity $\mathcal{H}_c(A)$ of its infinitesimal generator A. Roughly speaking, we shall prove here that vectors in $\mathcal{H}_c(A)$ tend weakly to zero under $\{U_t\}$ as $t \to \pm\infty$, but in general only when averaged over the parameter t in some sense. There are various ways of averaging that could be used, but we shall only consider the <u>Cesàro average</u> of a function $\varphi(t)$ defined as

$$\frac{1}{T}\int_0^T \varphi(t)dt.$$

We shall first give a simple lemma about the Cesàro limit, next deduce the basic theorem, the so-called *mean ergodic theorem*, and then deduce various consequences from it.

The first lemma states that, if a function $\varphi(t)$ has a limit as $t \to \infty$, then so does its Cesàro average as $T \to \infty$. The converse is of course not true; however, in some cases, if the Cesàro average converges, one can find a sequence $\{t_n\}$ such that $t_n \to \infty$ and such that $\varphi(t_n)$ converges to the same limit as the Cesàro average. One such case is treated in part (b) of the lemma.

<u>Lemma 4.18</u> : (a) Let $\varphi:\mathbb{R} \to \mathbb{R}$ be integrable and assume that $\lim\varphi(t) = a_\pm$ as $t \to \pm\infty$ respectively. Then

$$\lim_{T\to\infty} \pm\frac{1}{T}\int_0^{\pm T} \varphi(t)dt = a_\pm. \tag{4.65}$$

(b) Let $\varphi:\mathbb{R} \to [0,\infty)$ be such that (4.65) holds with $a_\pm = 0$. Then there are sequences $\{t_n^\pm\}$ such that $t_n^\pm \to \pm\infty$ and $\varphi(t_n^\pm) \to 0$ as $n \to \infty$.

(c) Let $\varphi:\mathbb{R} \to [0,\infty)$ be such that $\|\varphi\|_\infty \equiv \text{ess sup}_t\varphi(t) < \infty$. Then

$$\lim_{T\to\infty} \frac{1}{T}\int_0^T\varphi(t)dt = 0 \iff \lim_{T\to\infty} \frac{1}{T}\int_0^T|\varphi(t)|^2dt = 0.$$

<u>Proof</u> : We shall use the fact that, for $T > 0$,

$$\pm\frac{1}{T}\int_0^{\pm T}dt = 1. \tag{4.66}$$

(a) Set $\psi_\pm(t) = \varphi(t) - a_\pm$. Then one has for example for positive t :

$$|\frac{1}{T}\int_0^T \varphi(t)dt - a_+| = |\frac{1}{T}\int_0^T \psi_+(t)dt|$$

$$\leq \frac{1}{T}|\int_0^{T_1} \psi_+(t)dt| + \frac{1}{T}\int_{T_1}^T |\psi_+(t)|dt. \qquad (4.67)$$

Given $\varepsilon > 0$, one may choose T_1 such that $|\psi_+(t)| < \varepsilon/2$ for $t > T_1$, so that the second term on the r.h.s. is majorized by $\varepsilon/2$ (we assume $T > T_1$). As $T \to \infty$, the first term converges to zero, hence it is less than $\varepsilon/2$ provided that $T \geq T_0$. Thus the r.h.s. of (4.67) is less than ε if $T > \max(T_0, T_1)$.

(b) Let $\delta > 0$ and $s < \infty$. If $\varphi(t) \geq \delta$ for all $t \geq s$, one would have

$$\frac{1}{T}\int_0^T \varphi(t)dt \geq \delta(1 - \frac{s}{T}) \quad \text{for all} \quad T \geq s,$$

which contradicts the hypothesis that the Cesàro limit is zero. Hence there is a $t \geq s$ such that $\varphi(t) < \delta$.

It follows that, for each integer n, there is a number t_n such that $t_n \geq n$ and $\varphi(t_n) < 1/n$, which proves (b).

(c) We have on the one hand

$$\frac{1}{T}\int_0^T (\varphi(t))^2 dt \leq ||\varphi||_\infty \frac{1}{T}\int_0^T \varphi(t)dt.$$

On the other hand, by the Schwarz inequality,

$$\frac{1}{T}\int_0^T \varphi(t)dt \leq \frac{1}{T}[\int_0^T |\varphi(t)|^2 dt \cdot \int_0^T dt]^{1/2} = [\frac{1}{T}\int_0^T |\varphi(t)|^2 dt]^{1/2}. \quad \square$$

Let $\{U_t\}$ be an evolution group. Let N_0 and N_1 be defined as follows :

$$N_0 = \{f \in H | U_t f = f \quad \forall t \in \mathbb{R}\}, \qquad (4.68)$$

$$N_1 = \{g \in H | g = U_t f - f \text{ for some } f \in H \text{ and some } t \in \mathbb{R}\}, \qquad (4.69)$$

and let N be the subspace spanned by N_1. Then

Lemma 4.19 : (a) N_0 and N are subspaces each of which is invariant under $\{U_t\}$.

(b) N_0 and N are orthogonal complements of each other, i.e. $H = N_0 \oplus N$.

(c) N_0 is the null space of the infinitesimal generator A of $\{U_t\}$.

Proof : (a) This is very easy to prove. We omit the details.

(b) We have for all $f,g \in H$:

$$(U_t g - g, f) = (g, U_{-t} f - f). \tag{4.70}$$

Suppose that f is orthogonal to N. Then (4.70) implies that $U_{-t}f - f$ is orthogonal to H for each t. Hence $f \in N_0$, which proves that $H = N_0 \oplus N$.

(c) If $f \in N(A)$, then $f \in N_0$ by Corollary 4.5. Conversely, if $f \in N_0$, then $f \in D(A)$ and $Af = \theta$ by (4.6) and (4.7). Hence $N_0 = N(A)$. □

Proposition 4.20 (Mean Ergodic Theorem) : Let $\{U_t\}$ be an evolution group, A its infinitesimal generator and $E_0(A)$ the orthogonal projection with range N(A). Then, for each $f \in H$,

$$\operatorname*{s-lim}_{T\to\infty} \pm \frac{1}{T}\int_0^{\pm T} U_t f\, dt = E_0(A) f. \tag{4.71}$$

In particular, if $f \perp N(A)$, the above limit is zero.

Proof : (for the + sign) :

(i) Let $g \in N_1$, i.e. $g = U_s h - h$ for some $s \in \mathbb{R}$ and some $h \in H$. Then

$$\frac{1}{T}\int_0^T U_t g dt = \frac{1}{T}\int_s^{T+s} U_t h dt - \frac{1}{T}\int_0^T U_t h dt = -\frac{1}{T}\int_0^s U_t h dt + \frac{1}{T}\int_T^{T+s} U_t h dt.$$

Hence $\|\frac{1}{T}\int_0^T U_t g dt\| \le \frac{2|s|}{T} \|h\|$, which converges to zero as $T \to \infty$. Thus, for each g in the linear span $\mathcal{D}$ of $\mathcal{N}_1$:

$$\text{s-}\lim_{T\to\infty} \frac{1}{T}\int_0^T U_t g\,dt = 0. \qquad (4.72)$$

(ii) By Lemma 4.19, we may write $f = E_0(A)f + f_1$ with $f_1 \in \mathcal{N}$. Let $\varepsilon > 0$. Then there is a vector $g \in \mathcal{D}$ such that $\|f_1 - g\| < \varepsilon/2$, and

$$\|\frac{1}{T}\int_0^T U_t f\,dt - E_0(A)f\| = \|\frac{1}{T}\int_0^T U_t f_1\,dt\|$$

$$\le \|\frac{1}{T}\int_0^T U_t g\,dt\| + \|f_1 - g\|\frac{1}{T}\int_0^T dt.$$

The second term on the r.h.s. is less than $\varepsilon/2$ for each T, and the first term converges to zero by (4.72), hence it is less than $\varepsilon/2$ if $T \ge T_0$. This proves (4.71). □

Lemma 4.21 : Let $\{U_t\}$ be an evolution group, and assume that its infinitesimal generator A has purely continuous spectrum. Assume that $B \in \mathcal{B}_\infty$ and that $BU_t = U_t B$ for all $t \in \mathbb{R}$. Then $B = 0$. (In other words : The commutant of a self-adjoint operator A with purely continuous spectrum cannot contain any non-zero compact operator.)

Proof : (i) First assume that $B = B^*$. Let $\{\mathcal{M}_i\}$ be the eigensubspaces of B associated with its non-zero eigenvalues. By Proposition 3.12, each $\mathcal{M}_i$ is finite-dimensional and

$$\mathcal{H} = \bigoplus_i \mathcal{M}_i \oplus N(B).$$

By (4.34), we have in this decomposition of $\mathcal{H}$

$$U_t = \bigoplus_i U_{t;i} \oplus U_{t;o}.$$

As in the proof of Proposition 4.8, the restriction of A to $\mathcal{M}_i$ has pure point spectrum, which is impossible since, by assumption, $\mathcal{H}_p(A) = \{\theta\}$. Hence $\mathcal{M}_i = \{\theta\}$ for each i, and thus $\mathcal{H} = N(B)$, i.e. $B = 0$.

(ii) If $B \neq B^*$, then B^* also commutes with each U_t (cf. the proof of Proposition 4.6). Set $B_+ = B + B^*$, $B_- = i(B - B^*)$. Then B_+ and B_- are self-adjoint and commute with each U_t. Hence, by (i) : $B_+ = B_- = 0$. Now $B = \frac{1}{2}B_+ - \frac{1}{2}iB_-$, hence $B = 0$. □

<u>Lemma 4.22</u> : Let $\{U_t\}$ be an evolution group, A its infinitesimal generator. Assume that A has purely continuous spectrum. Then, for all $f,g \in \mathcal{H}$:

$$\lim_{T\to\pm\infty} \frac{1}{T}\int_0^{\pm T} |(g,U_t f)|^2 dt = 0. \tag{4.73}$$

<u>Proof</u> : This result will be obtained by applying the Mean Ergodic Theorem in the Hilbert space $\mathcal{B}_2 \equiv \mathcal{B}_2(\mathcal{H})$ to the following group $\{\mathcal{U}_t\}$:

$$\mathcal{U}_t B := U_t^* B U_t \ , \qquad B \in \mathcal{B}_2. \tag{4.74}$$

(i) Clearly $\mathcal{U}_t$ maps $\mathcal{B}_2$ into itself. We first check that it is an evolution group in $\mathcal{B}_2$, i.e. that it verifies (E1)-(E3). These follow from the corresponding properties for U_t.

$$\text{(E3)}\ \mathcal{U}_s \mathcal{U}_t B = U_s^*(U_t^* B U_t) U_s = U_{t+s}^* B U_{t+s} = \mathcal{U}_{s+t} B, \tag{4.75}$$

hence $\{\mathcal{U}_t\}$ has the group property.

(E1) Let t be fixed. Let $\{e_i\}$ be an orthonormal basis of $\mathcal{H}$, and let $f_i = U_t e_i$. Then $\{f_i\}$ is also an orthonormal basis of $\mathcal{H}$. Now by (3.8) we have for $B,C \in \mathcal{B}_2$:

$$\langle B,\mathcal{U}_t C\rangle = \sum_i (Be_i, U_t^* C U_t e_i) = \sum_i (U_t B U_t^* f_i, Cf_i)$$

$$= \sum_i (U_{-t}^* B U_{-t} f_i, Cf_i\rangle = \langle \mathcal{U}_{-t} B, C\rangle.$$

Hence $\mathcal{U}_t^* = \mathcal{U}_{-t}$, and the unitary of $\mathcal{U}_t$ now follows from this and (4.75) :

$$\mathcal{U}_t^*\mathcal{U}_t = \mathcal{U}_{-t}\mathcal{U}_t = \mathcal{U}_0 = \mathcal{I} = \mathcal{U}_t\mathcal{U}_t^*.$$

(E2) The strong continuity of $\mathcal{U}_t$ follows from Proposition 3.11 (b) .

$$||\mathcal{U}_{t+\tau}B - \mathcal{U}_tB||_{HS} = ||U_{t+\tau}^*BU_{t+\tau} - U_t^*BU_t||_{HS}$$

$$\leq ||(U_{t+\tau}^* - U_t^*)BU_t||_{HS} + ||U_{t+\tau}^*|| \, ||B(U_{t+\tau} - U_t)||_{HS} ,$$

which converges to zero as $\tau \to 0$ since $\{U_s^*\}$ is strongly continuous and $B, BU_t \in \mathcal{B}_2$.

(ii) By Lemma 4.19, we have an orthogonal decomposition of the Hilbert space $\mathcal{B}_2$, with respect to the evolution group $\{\mathcal{U}_t\}$, into $\mathcal{B}_2 = N_0 \oplus N$. N_0 is the set of all $B \in \mathcal{B}_2$ such that $\mathcal{U}_tB = B$, i.e. such that $U_t^*BU_t = B$, or, upon multiplication by U_t, such that $BU_t = U_tB$. Since A has purely continuous spectrum, Lemma 4.21 implies B = 0. Hence $N_0 = \{0\}$ and $\mathcal{B}_2 = N$. Thus, by Proposition 4.20,

$$\underset{T\to\infty}{\text{s-lim}} \frac{1}{T}\int_0^{\pm T} \mathcal{U}_tC\,dt = 0 \qquad (\text{in } \mathcal{B}_2),$$

for each $C \in \mathcal{B}_2$. Since strong convergence implies weak convergence, we have for all $B, C \in \mathcal{B}_2$:

$$\lim_{T\to\infty} \frac{1}{T}\int_0^{\pm T} \langle B, \mathcal{U}_tC\rangle dt = \lim_{T\to\infty} \frac{1}{T}\int_0^{\pm T} \sum_i (Be_i, U_t^*CU_te_i)dt = 0. \qquad (4.76)$$

Now let f,g be fixed non-zero vectors in H. We choose an orthonormal basis $\{e_i\}$ such that $e_1 = f/||f||$. Let us take in (4.76) for B and C the following finite rank operators :

$$Bh = (f,h)f, \qquad Ch = (g,h)g \qquad (h \in H).$$

We then get $Be_1 = ||f||f$, $Be_i = \theta$ for $i \geq 2$, $U_t^*CU_te_1 = ||f||^{-1}(g,U_tf)U_t^*g$, and insertion into (4.76) leads to

$$\lim_{T\to\infty} \frac{1}{T}\int_0^{\pm T} (f,U_t^*g)(g,U_tf)dt = \lim_{T\to\infty} \frac{1}{T}\int_0^{\pm T} |(g,U_tf)|^2 dt = 0. \qquad \square$$

Proposition 4.23 : Let $\{U_t\}$ be an evolution group, A its infinitesimal generator, and $f \in \mathcal{H}_c(A)$. Then, for each $g \in \mathcal{H}$:

$$\lim_{T\to\infty} \frac{1}{T}\int_0^{\pm T} |(g,U_t f)|^2 = 0, \tag{4.77}$$

$$\lim_{T\to\infty} \frac{1}{T}\int_0^{\pm T} |(g,U_t f)| = 0. \tag{4.78}$$

Proof : We write $\mathcal{H} = \mathcal{H}_p(A) \oplus \mathcal{H}_c(A)$ and $g = g_p \oplus g_c$ as in (2.7). By Proposition 4.6 (b), we have $(g,U_t f) = (g_c,U_{t,c}f)$. (4.77) now follows immediately from Lemma 4.22, since the restriction $U_{t,c}$ of U_t to $\mathcal{H}_c(A)$ is an evolution group and its infinitesimal generator A_c has purely continuous spectrum. (4.78) follows from (4.77) and Lemma 4.18 (c). □

Proposition 4.23 shows that the evolution $U_t f$ of each $f \in \mathcal{H}_c(A)$ converges weakly to zero on the average. One is then naturally led to introducing also the set of vectors that converge weakly to zero without any average. We therefore define

$$\mathcal{H}_w^{\pm}(A) = \{f \in \mathcal{H} | \lim_{t\to\pm\infty} (f,U_t f) = 0\}$$

and

$$\mathcal{H}_w(A) = \mathcal{H}_w^{+}(A) \cap \mathcal{H}_w^{-}(A). \tag{4.79}$$

We denote by $E_w(A)$ the projection with range $\mathcal{H}_w(A)$ and then have :

Proposition 4.24 : (a) $\mathcal{H}_w^{+}(A) = \mathcal{H}_w^{-}(A) = \mathcal{H}_w(A)$.

(b) If $f \in \mathcal{H}_w(A)$, then w-lim $U_t f = \theta$ as $t \to \pm\infty$.

(c) $\mathcal{H}_w(A)$ is a subspace of $\mathcal{H}_c(A)$, and it is invariant under U_s for each $s \in \mathbb{R}$.

(d) If $B \in \mathcal{B}(\mathcal{H})$ is such that $BU_s = U_s B$ for all $s \in \mathbb{R}$, then $BE_w(A) = E_w(A)B$.

Proof : (i) We have $(f,U_{-t}f) = (U^*_{-t}f,f) = (U_t f,f) = \overline{(f,U_t f)}$. If e.g. $f\in H^+_w(A)$, this implies that $f\in H^-_w(A)$, and vice versa. Hence $H^+_w(A) = H^-_w(A)$.

(ii) Let $s\in\mathbb{R}$, $f\in H_w(A)$. Then $(U_s f,U_t f) = (f,U_{t-s}f) \to 0$ as $t \to \pm\infty$. Thus $(g,U_t f) \to 0$ for each $g\in O(f)$, the orbit of f, hence for each g in the linear span $D(f)$ of $O(f)$.

Now if $h\in M(f)$, the subspace spanned by $O(f)$, and $\varepsilon > 0$, there is a $g_n\in D(f)$ such that $\|h - g_n\| \|f\| < \varepsilon/2$. Thus

$$|(h,U_t f)| = |(h - g_n,U_t f) + (g_n,U_t f)|$$

$$\leq \|h - g_n\| \|f\| + |(g_n,U_t f)| \leq \varepsilon \quad \text{if} \quad |t| \geq t_0(\varepsilon). \qquad (4.80)$$

Therefore $(h,U_t f) \to 0$ as $t \to \pm\infty$ for each $h\in M(f)$.

If $g\in H$, we may write $g = g_1 + g_2$ with $g_1\in M(f)$ and $g_2 \perp M(f)$. Then $(g,U_t f) = (g_1,U_t f) \to 0$ as $t \to \pm\infty$. This shows that w-lim $U_t f = \theta$ as $t \to \pm\infty$ for each $f\in H_w(A)$.

(iii) If $f\in H_w(A)$, then $U_s f\in H_w(A)$, since $(U_s f,U_t U_s f) = (f,U_t f)$. Thus $H_w(A)$ is invariant under U_s.

(iv) Let $f,g\in H_w(A)$, $\alpha\in\mathbb{C}$. Then

$$(f+\alpha g,U_t(f+\alpha g)) = (f,U_t f) + \alpha(f,U_t g) + \bar{\alpha}(g,U_t f) + |\alpha|^2(g,U_t g),$$

which converges to zero as $|t| \to \infty$ by (ii). Therefore $H_w(A)$ is a linear manifold. The proof that $H_w(A)$ is strongly closed (i.e. that is a subspace) is similar to the argument in (4.80).

(v) Let g be an eigenvector of A, say $Ag = \lambda g$, and $f\in H_w(A)$. Then by Corollary 4.5,

$$(g,f) = (e^{-i\lambda t}U^*_t g,f) = e^{i\lambda t}(g,U_t f) \to 0 \quad \text{as} \quad t \to \pm\infty.$$

Hence $f \perp g$. This shows that $H_w(A) \perp H_p(A)$, i.e. that $H_w(A) \subseteq H_c(A)$.

(vi) Let $f \in H_w(A)$. Then, if B commutes with U_t :

$$(Bf, U_t Bf) = (B^*Bf, U_t f) \to 0 \quad \text{as} \quad t \to \pm\infty.$$

Hence $Bf \in H_w(A)$, i.e. $BE_w = E_w BE_w$. Since B^* also commutes with each U_t, we obtain similarly that $B^*E_w = E_w B^*E_w$, or, upon taking the adjoint : $E_w B = E_w BE_w$. By combining this with $BE_w = E_w BE_w$, we arrive at (d). □

From spectral theory it is known that the so-called subspace of absolute continuity $H_{ac}(A)$ of A is contained in $H_w(A)$. This subspace can also be introduced in terms of the evolution group $\{U_t\}$ associated with A :

$$H_{ac}(A) := \text{subspace spanned by } \{f \in H \mid \int_{\mathbb{R}} |(f, U_t f)|^2 dt < \infty\}. \quad (4.81)$$

A proof that this definition of $H_{ac}(A)$ is equivalent to that in terms of the spectral family of A may be found in [1]. A heuristic description of the latter definition and a verification of the fact that $H_{ac}(A) \subseteq H_w(A)$ will be given at the end of this section (Remark 4.31 and Lemma 4.30).

Proposition 4.25 : Let $\{U_t\}$ be an evolution group, A its infinitesimal generator and B an operator in $B(H)$. Assume there is an operator $C \in B(H)$ such that (i) $CU_t = U_t C$ for all $t \in \mathbb{R}$, (ii) $R(C)$ is dense in H, (iii) $BC \in B_\infty$. Then, for each $f \in H_c(A)$:

$$\lim_{T\to\infty} \frac{1}{T}\int_0^{\pm T} \|BU_t f\|^2 dt = 0, \quad (4.82)$$

$$\lim_{T\to\infty} \frac{1}{T}\int_0^{\pm T} \|BU_t f\| dt = 0. \quad (4.83)$$

In addition, for each $f \in H_w(A)$:

$$\lim_{t\to\pm\infty} ||BU_t f|| = 0. \qquad (4.84)$$

In particular, (4.82)-(4.84) hold if $B\in\mathcal{B}_\infty$ or if $B(A-zI)^{-1}\in\mathcal{B}_\infty$ for some $z\in\rho(A)$.

Proof : Let $\varepsilon > 0$ and assume $B \neq 0$. By (ii), there is a vector $f_0\in\mathcal{H}$ such that $||f - Cf_0||^2 < \varepsilon(9||B||^2)^{-1}$. By (i) and (4.32) : $CE_c(A) = E_c(A)C$. Let $h = E_c(A)f_0$. Then

$$||f - Ch||^2 = ||E_c(A)f - E_c(A)Cf_0||^2 \leq ||f - Cf_0||^2 < \varepsilon(9||B||^2)^{-1}.$$

If $h = \theta$, then the integral in (4.82) is less than $\varepsilon T||B||^{-2}$ for all T. If $h \neq \theta$, choose a finite rank operator D of the form (3.20) such that $||BC - D||^2 < \varepsilon(9||h||^2)^{-1}$. Then, by (1.13) :

$$||BU_t f||^2 \leq 3||BU_t(f - Ch)||^2 + 3||(BC - D)U_t h||^2 + 3||DU_t h||^2$$

$$< \varepsilon/3 + \varepsilon/3 + 3||\sum_{k=1}^N (g_k, U_t h)h_k||^2 \leq 2\varepsilon/3 + 3N\sum_{k=1}^N |(g_k, U_t h)|^2 ||h_k||^2.$$

Hence, using also (4.66), one obtains that

$$\pm\frac{1}{T}\int_0^{\pm T} ||BU_t f||^2 dt \leq 2\varepsilon/3 + 3N\sum_{k=1}^N ||h_k||^2 (\pm\frac{1}{T})\int_0^{\pm T} |(g_k, U_t h)|^2 dt.$$

The last term is a finite sum of terms each of which converges to zero as $T \to \infty$ by Proposition 4.23; hence it is less than $\varepsilon/3$ provided $|T| \geq T_0$. This proves (4.82). (4.83) is obtained from (4.82) and Lemma 4.18 (c). Similarly one obtains (4.84) by using Proposition 4.24 instead of Proposition 4.23.

If $B\in\mathcal{B}_\infty$, we may take $C = I$, and if $B(A - zI)^{-1}\in\mathcal{B}_\infty$, we choose $C = (A - zI)^{-1}$; in the second case (i) follows for example from Corollary 4.2 (if $\text{Im}z \neq 0$) and (ii) from Proposition 2.9 (a). □

Lemma 4.26 : Assume that $\tilde{\varphi}\in L^1(\mathbb{R})$. Let $f\in\mathcal{H}_c(A)$. Then

$$\text{s-}\lim_{T\to\infty} \frac{1}{T}\int_0^{\pm T} d\tau \int_{-\infty}^{\infty} ds\tilde{\varphi}(s) U^*_{\tau s} f = \theta. \tag{4.85}$$

Proof : By Fubini's Theorem (applied e.g. to the scalar product with any vector $g\in H$), we may change the order of integration in (4.85), so that

$$\|\frac{1}{T}\int_0^{\pm T} d\tau \int_{-\infty}^{\infty} ds\tilde{\varphi}(s) U^*_{\tau s} f\| = \|\int_{-\infty}^{\infty} ds\tilde{\varphi}(s)\frac{1}{T}\int_0^{\pm T} U^*_{\tau s} f d\tau\|$$

$$\leq \int_{-\infty}^{\infty} ds\,|\tilde{\varphi}(s)|\,\|\frac{1}{T}\int_0^{\pm T} U^*_{\tau s} f d\tau\|, \tag{4.86}$$

where we have also used (1.57). Now the integrand of the integral over ds is majorized, uniformly in $T\in\mathbb{R}$, by the L^1-function $|\tilde{\varphi}(s)|\,\|f\|$, and it converges to zero for each $s \neq 0$ for which $|\tilde{\varphi}(s)| < \infty$, by the Mean Ergodic Theorem. Hence, by the Lebesgue dominated convergence theorem, each double integral in (4.86) converges to zero as $T \to \infty$, which proves (4.85). □

We use this lemma to prove a result that will be useful in Chapter 5. We set

$$C_{00}^{\infty}(\mathbb{R}) := C_0^{\infty}(\mathbb{R}\backslash\{0\}). \tag{4.87}$$

Thus $C_{00}^{\infty}(\mathbb{R})$ is the set of all infinitely differentiable functions from $\mathbb{R}$ to $\mathbb{C}$ each of which vanishes in some neighbourhood of $\lambda = 0$ and outside some finite interval.

Proposition 4.27 : Let A be self-adjoint, M a subspace of $H_c(A)$ invariant under U_t and D a total subset of M. Then the set $\{\varphi(A)D|\varphi\in C_{00}^{\infty}(\mathbb{R})\}$ is total in M.

Proof : Assume that $g\in M$ and that $(g,\varphi(A)f) = 0$ for all $f\in D$ and all $\varphi\in C_{00}^{\infty}(\mathbb{R})$. We must show that $g = \theta$, cf. Lemma 1.5. For this, let $\varphi\in C_0^{\infty}(\mathbb{R})$ be as in (4.52), and define φ_τ by

$$\varphi_\tau(\lambda) = \varphi(\tau\lambda) - \varphi(\lambda/\tau) \qquad (\tau > 0). \tag{4.88}$$

Then $\widetilde{\varphi}_\tau(t) = \frac{1}{\tau}\widetilde{\varphi}(t/\tau) - \tau\widetilde{\varphi}(\tau t)$,

and consequently

$$\varphi_\tau(A)f = (2\pi)^{-1/2} \int_{-\infty}^{\infty} [\frac{1}{\tau}\widetilde{\varphi}(t/\tau) - \tau\widetilde{\varphi}(\tau t)]U_t^* f\, dt$$

$$= (2\pi)^{-1/2} \int_{-\infty}^{\infty} \widetilde{\varphi}(s)U_{\tau s}^* f\, ds - (2\pi)^{-1/2} \int_{-\infty}^{\infty}\widetilde{\varphi}(s)U_{s/\tau}^* f\, ds.$$

Since $\varphi_\tau \in C_{oo}^\infty(\mathbb{R})$, we have $(g,\varphi_\tau(A)f) = 0$ for each $\tau > 0$ and each $f \in \mathcal{D}$. Hence, for each $T > 0$ and each $f \in \mathcal{D}$:

$$0 = \frac{1}{T} \int_0^T (g,\varphi_\tau(A)f)d\tau = (2\pi)^{-1/2} \frac{1}{T} \int_0^T d\tau \int_{-\infty}^{\infty} ds\widetilde{\varphi}(s)(g,U_{\tau s}^* f)$$

$$- (2\pi)^{-1/2} \frac{1}{T} \int_0^T d\tau \int_{-\infty}^{\infty} ds\widetilde{\varphi}(s)(g,U_{s/\tau}^* f).$$

As $T \to \infty$, the first term on the r.h.s. converges to zero by Lemma 4.26; on the other hand the second term converges to $-\varphi(0)(g,f)$, which may easily be derived by first using Lemma 4.10 and then part (a) of Lemma 4.18. Thus we have proved that $0 = -\varphi(0)(g,f) = -(g,f)$ for each $f \in \mathcal{D}$. Since $\mathcal{D}$ is total in M by hypothesis, we must have $g = \theta$. This completes the proof. □

Remark 4.28 : Proposition 4.27 can also be proven for the case where $C_{oo}^\infty(\mathbb{R})$ is replaced by certain other subsets of $C_o^\infty(\mathbb{R})$, for instance by $C_o^\infty(\mathbb{R}\backslash\Gamma)$, where Γ is a finite set or a closed countable set. However, for our later applications, it is sufficient to know this result for $C_{oo}^\infty(\mathbb{R})$.

Next we give another consequence of the preceding ergodic properties. This result will not be used in later chapters.

Proposition 4.29 : Let $\{U_t\}$ be an evolution group and $U_{t,c}$ its restriction to the continuous subspace $H_c(A)$ of its infinitesimal generator A. Then there are two sequences $\{t_n^\pm\}$ such that $t_n^+ \to +\infty$,

$t_n^- \to -\infty$ and $\text{w-lim}\, U_{t_n^\pm, c} = 0$ as $n \to \infty$.

Proof (for the + sign) :

(i) Let $\{e_k\}$ be an orthonormal basis of $H_c(A)$. Define a linear operator B by $Bf = \sum_{k=1}^{\infty} \frac{1}{k}(e_k, f)e_k$. Then

$$||Bf||^2 = \sum_{k=1}^{\infty} k^{-2}|(e_k, f)|^2 \leq ||f||^2 \sum_{k=1}^{\infty} k^{-2},$$

which shows that $B \in \mathcal{B}(H)$. In fact B is a Hilbert-Schmidt operator, since

$$||B||_{HS}^2 = \sum_{k=1}^{\infty} ||Be_k||^2 = \sum_{k=1}^{\infty} k^{-2} < \infty.$$

Now consider the quantity

$$\frac{1}{T}\int_0^T \sum_{r=1}^{\infty} r^{-2}||BU_t e_r||^2 dt = \sum_{r=1}^{\infty} r^{-2} \frac{1}{T}\int_0^T ||BU_t e_r||^2 dt.$$

As $T \to \infty$, each term in this sum converges to zero, by Proposition 4.25. Also, each term is majorized, uniformly in $T \geq 0$, by $||B||^2 r^{-2}$. Therefore, by the discrete version of the Lebesgue dominated convergence theorem,

$$\lim_{T\to\infty} \frac{1}{T}\int_0^T \sum_{r=1}^{\infty} r^{-2}||BU_t e_r||^2 dt = 0.$$

By Lemma 4.18 (b), there is a sequence $\{t_n\}$ such that $t_n \to +\infty$ and

$$\lim_{n\to\infty} \sum_{r=1}^{\infty} r^{-2}||BU_{t_n} e_r||^2 = 0. \tag{4.89}$$

Now

$$||BU_{t_n} e_r||^2 = \sum_{k=1}^{\infty} k^{-2}|(e_k, U_{t_n} e_r)|^2,$$

so that by (4.89)

$$\lim_{n\to\infty}(e_k, U_{t_n} e_r) = 0 \qquad \forall\, k,r = 1,2,\ldots \,. \tag{4.90}$$

(ii) Let $\mathcal{D}$ be the linear span of $\{e_k\}$. $\mathcal{D}$ is dense in $\mathcal{H}_c(A)$ and, by (4.90),

$$\lim_{n\to\infty}(f, U_{t_n} g) = 0 \qquad \forall\, f,g\in\mathcal{D}. \tag{4.91}$$

Let $h_1, h_2 \in \mathcal{H}_c(A)$ be non-zero vectors and $\varepsilon > 0$. Choose $f,g\in\mathcal{D}$ such that $\|h_1 - f\| < \varepsilon(3\|h_2\|)^{-1}$ and $\|h_2 - g\| < \varepsilon(3\|f\|)^{-1}$. Then

$$|(h_1, U_{t_n} h_2)| \le |(h_1 - f, U_{t_n} h_2)| + |(U^*_{t_n} f, h_2 - g)| + |(f, U_{t_n} g)|$$

$$< \varepsilon/3 + \varepsilon/3 + |(f, U_{t_n} g)|,$$

which, by (4.91), is less than ε provided $n \ge n_o$, where n_o depends on h_1, h_2 and ε. This proves that $U_{t_n,c}$ converges weakly to zero as $n \to \infty$. □

We end this section with a few considerations regarding spectral subspaces of a self-adjoint operator A. We first prove that each vector in the subspace of absolute continuity of A, defined in (4.81), converges weakly to zero under the associated evolution group :

Lemma 4.30 : Let A be self-adjoint and let $\{U_t\}$ be the associated evolution group. Then

$$\mathcal{H}_{ac}(A) \subseteq \mathcal{H}_w(A).$$

Proof : (i) For each $f\in\mathcal{H}$, the function $\varphi(t) := (f, U_t f)$ is uniformly continuous in t, i.e. given any $\varepsilon > 0$ there is a $\tau_o > 0$ (depending only on ε) such that $|\varphi(t+\tau) - \varphi(t)| < \varepsilon$ whenever $|\tau| \le \tau_o$. This is so because U_t is strongly continuous at $t = 0$:

$$|(f, (U_{t+\tau} - U_t) f)| = |(U_t f, (U_\tau - I) f)| \le \|f\|\,\|(U_\tau - I) f\|.$$

(ii) Since $H_w(A)$ is a subspace, the inclusion $H_{ac}(A) \subseteq H_w(A)$ will be established as soon as we have shown that each f satisfying $\int |(f,U_t,f)|^2 dt < \infty$ belongs to $H_w(A) = H_w^+(A)$. But this follows immediately from Lemma A.5, which states that the finiteness of the preceding integral and the uniform continuity of $(f,U_t f)$ together imply that $(f,U_t f) \to 0$ as $t \to +\infty$. □

Remark 4.31 : Using the Fourier transforms with respect to the variable t of the functions $(f,U_t f)$, it is possible to introduce a further subdivision of the Hilbert space $H_c(A)$. One can show that $(f,U_t f)$ may be represented in the form

$$(f,U_t f) = \int_{-\infty}^{\infty} e^{-i\lambda t} d\mu_f(\lambda),$$

where μ_f is a function from $\mathbb{R}$ to $[0,\infty)$ which depends on f and on the operator A and satisfies

(α) $\mu_f(\lambda) \geq \mu_f(\lambda')$ if $\lambda \geq \lambda'$,

(β) $\mu_f(-\infty) := \lim_{\lambda\to-\infty} \mu_f(\lambda) = 0$,

(γ) $\int_{-\infty}^{\infty} d\mu_f(\lambda) = \mu_f(+\infty) = ||f||^2 < \infty$,

(δ) $\lim_{\varepsilon\to+0} \mu_f(\lambda + \varepsilon) = \mu_f(\lambda) \quad \forall \lambda \in \mathbb{R}$.

This is a special case of a theorem due to Bochner (see e.g. [B;§21], [HO;§4]).

A vector $f \in H$ is said to belong to the subspace of absolute continuity $H_{ac}(A)$ of A if the measure determined by μ_f is absolutely continuous with respect to Lebesgue measure on $\mathbb{R}$, i.e. if $\mu_f(\Delta) = 0$ for every Borel subset Δ of $\mathbb{R}$ having Lebesgue measure $|\Delta| = 0$. f is said to belong to the subspace of singularity $H_s(A)$

of A if μ_f is singular with respect to Lebesgue measure, i.e. if there is a Borel set Δ_0 of zero Lebesgue measure, $|\Delta_0| = 0$, such that $\mu_f(\Delta) = \mu_f(\Delta \cap \Delta_0)$ for each Borel set Δ of $\mathbb{R}$. It can be shown that $H_{ac}(A)$ and $H_s(A)$ are invariant under U_t and orthogonal complements of each other, i.e. that $H = H_{ac}(A) \oplus H_s(A)$ and also that $A = A_{ac} \oplus A_s$ as in (2.8).

If f is such that $(f,U_tf) \in L^p(\mathbb{R})$ for some $p \in [1,2]$, then $|(f,U_tf)|^2 \leq c|(f,U_tf)|^p$ for some constant $c < \infty$ depending only on f. Thus (f,U_tf) is also in $L^2(\mathbb{R})$, hence f is in $H_{ac}(A)$ according to the definition (4.81). Thus $\varphi(t) := (f,U_tf)$ is the Fourier transform of a function $\hat{\varphi}$ which is in $L^2(\mathbb{R})$, hence locally in $L^1(\mathbb{R})$:

$$(f,U_tf) = \int_{-\infty}^{\infty} e^{-i\lambda t}\hat{\varphi}(\lambda)d\lambda.$$

It is then clear (and can be rigorously shown) that $\hat{\varphi} \in L^1(\mathbb{R})$ and $d\mu_f(\lambda) = \hat{\varphi}(\lambda)d\lambda$. Hence in this case the measure μ_f is absolutely continuous with respect to Lebesgue measure, so that $f \in H_{ac}(A)$ also by the definition of $H_{ac}(A)$ via Fourier transformation.

Suppose now that f is an eigenvector of A, say $Af = \alpha f$. Then $(f,U_tf) = \exp(-i\alpha t)\|f\|^2$. In this case the function μ_f must be as follows : $\mu_f(\lambda) = 0$ for $\lambda < \alpha$, $\mu_f(\lambda) = \|f\|^2$ for $\lambda \geq \alpha$. Thus the support of the measure μ_f consists of just one point $\Delta_0 = \{\alpha\}$, a set of Lebesgue measure zero. This shows that each eigenvector of A belongs to $H_s(A)$. Hence $H_p(A) \subseteq H_s(A)$ and consequently $H_{ac}(A) \subseteq H_c(A)$. (We have already pointed out that the definition of $H_{ac}(A)$ given here is equivalent to that given in (4.81), see [1]).

Apart from $H_p(A)$, the subspace of singularity $H_s(A)$ may also contain a part of $H_c(A)$, in which case $H_{ac}(A)$ is strictly smaller than $H_c(A)$. The intersection of $H_c(A)$ and $H_s(A)$ is called the <u>sub-</u>

space of singular continuity of A and denoted by $\mathcal{H}_{sc}(A)$. One then has the following decomposition of $\mathcal{H}$ and A :

$$\mathcal{H} = \mathcal{H}_{ac}(A) \oplus \mathcal{H}_{sc}(A) \oplus \mathcal{H}_{p}(A),$$

$$A = A_{ac} \oplus A_{sc} \oplus A_{p} ,$$

where each of the three operators on the r.h.s. is self-adjoint in its respective subspace. The spectrum of A_{ac}, as an operator in $\mathcal{H}_{ac}(A)$, is called the absolutely continuous spectrum $\sigma_{ac}(A)$ of A; the spectrum of A_{sc}, as an operator in $\mathcal{H}_{sc}(A)$, is called the singularly continuous spectrum $\sigma_{sc}(A)$ of A. If e.g. $\mathcal{H}_{sc}(A) = \{\theta\}$, then A is said to have no singularly continuous spectrum, i.e. $\sigma_{sc}(A) = \emptyset$. Clearly $\sigma_{c}(A) = \sigma_{ac}(A) \cup \sigma_{sc}(A)$.

If $f \in \mathcal{H}_{sc}(A)$, then $\mu_f(\lambda)$ is a continuous function the derivative of which is zero for almost all λ. Such functions can be constructed by using the Cantor measure. Similarly, the Cantor measure allows one to obtain examples in which there exist vectors in $\mathcal{H}_c(A)$ such that $(f,U_t f)$ does not converge to zero as $t \to \pm\infty$, i.e. such that $\mathcal{H}_w(A)$ is strictly smaller than $\mathcal{H}_c(A)$. However $\mathcal{H}_w(A)$ may be strictly larger than $\mathcal{H}_{ac}(A)$, so that one has in general

$$\mathcal{H}_{ac}(A) \subseteq \mathcal{H}_{w}(A) \subseteq \mathcal{H}_{c}(A) , \qquad (4.92)$$

with strict inclusion possible in both places. For examples relative to these points, we refer to [Z;§XII.10/11], [2]. Schrödinger operators with singular continuous spectrum are constructed in [3].

4.4 The Schrödinger Free Evolution Group

In Example 2.15 we introduced the so-called Schrödinger free Hamiltonian H_o, the self-adjoint operator determined by the negative Laplacian in $L^2(\mathbb{R}^\nu)$. The associated evolution group will be denoted by $\{U_t^o\}$ and called the Schrödinger free evolution group. In the present section we give a few simple properties of the operators U_t^o.

Proposition 4.32 : In $\tilde{L}^2(\mathbb{R}^\nu)$, U_t^o is given as

$$(FU_t^o f)(\underline{k}) = \exp(-i|\underline{k}|^2 t)\tilde{f}(\underline{k}). \tag{4.93}$$

Proof : We take (4.93) as the definition of U_t^o. We must then show that the family $\{U_t^o\}_{t\in\mathbb{R}}$ forms an evolution group and that its infinitesimal generator is H_o. Formally this is clear, but a rigorous proof requires some care.

(i) Using the properties of the exponential function, one finds that $||U_t^o f||^2 = ||f||^2$, $U_t^o U_s^o = U_{t+s}^o$ and $U_t^{o*} = U_{-t}^o$. Hence $U_t^o \in \mathcal{B}(H)$, the family $\{U_t^o\}$ has the group property, and U_t^o is unitary : $U_t^{o*}U_t^o = U_{-t}^o U_t^o = I = U_t^o U_t^{o*}$. Furthermore

$$||U_t^o f - U_s^o f||^2 = \int |e^{i\underline{k}^2 t} - e^{i\underline{k}^2 s}|^2 |\tilde{f}(\underline{k})|^2 d^\nu k.$$

The integrand converges to zero for almost all $\underline{k}\in\mathbb{R}^\nu$ as $s \to t$, and it is majorized by the L^1-function $4|\tilde{f}(\underline{k})|^2$. Hence, by the Lebesgue dominated convergence theorem, the integral converges to zero as $s \to t$, which establishes the strong continuity of $\{U_t^o\}$.

(ii) To verify that the infinitesimal generator of $\{U_t^o\}$ is the self-adjoint operator H_o, it suffices to show that $id/dt\, U_t^o f|_{t=0} = H_o f$ for each $f\in D(H_o)$. Now

$$||\tfrac{i}{t}(U_t^o - I)f - H_o f||^2 = \int |\tfrac{i}{t}(e^{-i\underline{k}^2 t} - 1) - \underline{k}^2|^2 |\tilde{f}(\underline{k})|^2 d^\nu k. \tag{4.94}$$

Again the integrand converges to zero for almost all $\underline{k}\in\mathbb{R}^\nu$ as $t \to 0$. Since

$$|\tfrac{i}{t}(e^{-i\underline{k}^2 t} - 1) - \underline{k}^2| = |\int_0^{\underline{k}^2} (e^{-itx} - 1)dx| \le 2\underline{k}^2,$$

this integrand is also majorized, uniformly in $t\in\mathbb{R}$, by the function $4|\underline{k}|^4|\tilde{f}(\underline{k})|^2$, which is in $L^1(\mathbb{R}^\nu)$ provided that $f\in D(H_o)$, cf. (2.27). Hence, for $f\in D(H_o)$, the limit as $t\to 0$ in (4.94) is zero.□

Proposition 4.33 : In $L^2(\mathbb{R}^\nu)$, U_t^o is an integral operator given by $(t\neq 0)$

$$(U_t^o f)(\underline{x}) = (4\pi it)^{-\nu/2}\int \exp[i|\underline{x}-\underline{y}|^2/(4t)]f(\underline{y})d^\nu y, \tag{4.95}$$

where the branch of the square root is such that

$$\left(\frac{4\pi it}{|4\pi t|}\right)^{-\nu/2} = \begin{cases} \exp(-i\nu\pi/4) & \text{if } t>0 \\ \exp(+i\nu\pi/4) & \text{if } t<0. \end{cases}$$

Remark 4.34 : If $f\in L^1(\mathbb{R}^\nu)\cap L^2(\mathbb{R}^\nu)$, the integral in (4.95) exists for each $\underline{x}\in\mathbb{R}^\nu$. For general $f\in L^2(\mathbb{R}^\nu)$, (4.95) has to be interpreted as a limit in the mean (see page 10).

Proof : We only give a formal proof and refer to [AJS, Lemma 3.12] for a rigorous derivation. We set

$$M_t(a) := (2\pi)^{-1}\int_{-\infty}^{\infty} e^{ika}e^{-ik^2t}dk = (4\pi it)^{-1/2}\exp(ia^2/4t)$$

and

$$N_t(\underline{x},\underline{y}) := (2\pi)^{-\nu}\int e^{i\underline{k}\cdot(\underline{x}-\underline{y})}e^{-i\underline{k}^2t}d^\nu k = \prod_{r=1}^{\nu} M_t(x_r-y_r). \tag{4.96}$$

Interchanging freely the order of integration, we obtain from (4.93) and (1.21) that

$$(U_t^o f)(\underline{x}) = (2\pi)^{-\nu/2}\int e^{i\underline{k}\cdot\underline{x}}e^{-i\underline{k}^2t}\tilde{f}(\underline{k})d^\nu k$$

$$= (2\pi)^{-\nu}\int d^\nu k\, e^{i\underline{k}\cdot\underline{x}}e^{-i\underline{k}^2t}\int e^{-i\underline{k}\cdot\underline{y}}f(\underline{y})d^\nu y = \int N_t(\underline{x},\underline{y})f(\underline{y})d^\nu y.$$

Insertion of (4.96) into this expression gives (4.95). □

Proposition 4.35 : The Schrödinger free Hamiltonian H_o is spectrally absolutely continuous, i.e. one has

$$\mathcal{H}_{ac}(H_o) = \mathcal{H}_w(H_o) = L^2(\mathbb{R}^\nu).$$

Proof : Let $f \in L^2(\mathbb{R}^\nu) \cap L^1(\mathbb{R}^\nu)$. We have $|(f, U_t^o f)| \leq \|f\|^2$. On the other hand, by (4.95),

$$|(f, U_t^o f)| = |4\pi t|^{-\nu/2} |\iint \overline{f(\underline{x})} f(\underline{y}) \exp[i|\underline{x} - \underline{y}|^2/(4t)] d^\nu x \, d^\nu y$$

$$\leq (4\pi|t|)^{-\nu/2} \|f\|_1^2.$$

Hence $|(f, U_t^o f)| \leq \min\{\|f\|^2, |t|^{-\nu/2}\|f\|_1^2\}$, which is in $L^2(\mathbb{R})$ as a function of t if $\nu \geq 2$. Thus, by the definition (4.81), $\mathcal{H}_{ac}(H_o)$ contains the dense set $L^2(\mathbb{R}^\nu) \cap L^1(\mathbb{R}^\nu)$, so that $\mathcal{H}_{ac}(H_o) = L^2(\mathbb{R}^\nu)$ if $\nu \geq 2$. The second identity now follows from Lemma 4.30. For $\nu = 1$, see Problem 4.36. □

Problem 4.36 : Use (4.93) directly to prove that $\mathcal{H}_{ac}(H_o) = L^2(\mathbb{R}^\nu)$ for each $\nu \geq 1$.

Proposition 4.37 : Let $\mathcal{H} = L^2(\mathbb{R}^\nu)$. For $t \neq 0$, define an operator Z_t by

$$(Z_t f)(\underline{x}) = \exp(i\underline{x}^2/4t) f(\underline{x}). \qquad (4.97)$$

Then Z_t is unitary and satisfies

(a) $$\underset{t\to\pm\infty}{\text{s-lim}}\, Z_t = I, \qquad (4.98)$$

(b) $$U_t^{o*} \varphi(\underline{Q}) U_t^o = Z_t^* \varphi(2t\underline{P}) Z_t \qquad \forall \varphi \in L^\infty(\mathbb{R}^\nu). \qquad (4.99)$$

Proof : (i) Clearly $Z_t^* = Z_{-t}$ and $Z_{-t}Z_t = Z_t Z_{-t} = I$, so that Z_t is unitary. (4.98) is obtained by using the Lebesgue dominated convergence theorem as in the proof of Proposition 4.32 and noticing that $(Z_t f)(\underline{x}) \to f(\underline{x})$ as $t \to \pm\infty$ for each $\underline{x}$ and $|(Z_t f)(\underline{x}) - f(\underline{x})|^2 \le 4|f(\underline{x})|^2 \in L^1(\mathbb{R}^\nu)$.

(ii) Let $f \in S(\mathbb{R}^\nu)$. We use (4.95) to express $U_t^{o*}\varphi(\underline{Q})U_t^o f$ and then make the change of variables $\underline{z} \mapsto \underline{k} = \underline{z}/2t$:

$$(U_t^{o*}\varphi(\underline{Q})U_t^o f)(\underline{x}) = \frac{1}{|4\pi t|^\nu} \int d^\nu z \, e^{-i|\underline{x}-\underline{z}|^2/4t}\varphi(\underline{z}) \int e^{i|\underline{z}-\underline{y}|^2/4t} f(\underline{y}) d^\nu y$$

$$= \frac{1}{(2\pi)^\nu} e^{-i\underline{x}^2/4t} \int d^\nu k \, e^{i\underline{k}\cdot\underline{x}}\varphi(2t\underline{k}) \int e^{-i\underline{k}\cdot\underline{y}} e^{i\underline{y}^2/4t} f(\underline{y}) d^\nu y.$$

The integrals exist pointwise since $f \in S(\mathbb{R}^\nu)$ and $\varphi \in L^\infty(\mathbb{R}^\nu)$. Now it is easily seen that the last expression is just $[Z_t^* \varphi(2t\underline{P})Z_t f](\underline{x})$. Hence the identity (4.99) holds on $S(\mathbb{R}^\nu)$, and since all operators occuring in it are bounded, it holds on all of $L^2(\mathbb{R}^\nu)$ by continuity (apply Lemma 1.8 with $A_n = U_t^{o*}\varphi(\underline{Q})U_t$ for each n, $A = Z_t^* \varphi(2t\underline{P})Z_t$ and $\mathcal{D}_o = S(\mathbb{R}^\nu)$). □

Problem 4.38 : Let $\varphi: \mathbb{R} \to \mathbb{C}$ be the inverse Fourier transform of a function $\tilde{\varphi} \in L^1(\mathbb{R})$. Then $\varphi(H_o)$ is the multiplication operator in $\tilde{L}^2(\mathbb{R}^\nu)$ by $\varphi(\underline{k}^2)$. In addition $\|\varphi(H_o)\| = \|\varphi\|_\infty$.

CHAPTER 5 : ASYMPTOTIC PROPERTIES OF EVOLUTION GROUPS

In this chapter we prove a number of results on the behaviour of $U_t f$ as the parameter t tends to $+\infty$ or $-\infty$.

To classify vectors according to their asymptotic behaviour under U_t, one must consider some additional mathematical structure apart from the evolution group $\{U_t\}$. In Section 5.1 this will be a family of "localizing operators". In $L^2(\mathbb{R}^\nu)$, these operators are just the spectral measure of the ν-component position operator introduced in Example 2.13. We shall classify vectors according to how they behave in configuration space $\mathbb{R}^\nu$ as a function of t, and in particular define the so-called bound states and scattering states.

Roughly speaking, the scattering states are vectors whose probability of being localized in any given bounded region of configuration space tends to zero as $t \to \pm\infty$. In Section 5.2 we consider as additional mathematical structure a second evolution group $\{U_t^o\}$ and ask whether it is possible to describe the asymptotic behaviour of the scattering states of the first group $\{U_t\}$ in terms of the second group $\{U_t^o\}$. Physically this corresponds to asking whether the asymptotic condition is satisfied with respect to a group $\{U_t^o\}$ describing a "non-interacting" system; mathematically it is one possible way of asking whether the infinitesimal generator A of $\{U_t\}$ (or its continuous part A_c) is unitarily equivalent to a simpler operator, the infinitesimal generator A^o of U_t^o. (In our applications A^o will just be the Schrödinger free Hamiltonian H_o of Example 2.15.)

The asymptotic comparison of $\{U_t\}$ and $\{U_t^o\}$ can be done in terms of the wave operators. These will be defined in Section 5.2, where we shall also establish their basic properties. In Section 5.3 we give an abstract criterion for existence and completeness of the wave operators, and in Section 5.4 we apply it to the concrete case of evolution groups determined by Schrödinger Hamiltonians.

5.1 Bound States, Scattering States and Absorbed States

In this section we consider the concrete situation where $H = L^2(\mathbb{R}^\nu)$, although some of the material could also be presented abstractly. We denote by B_r the ball of radius $r > 0$ in $\mathbb{R}^\nu$ centered at the origin :

$$B_r = \{\underline{x}\in\mathbb{R}^\nu \,\big|\, |\underline{x}| \le r\}. \tag{5.1}$$

Let F_r be the orthogonal projection in $L^2(\mathbb{R}^\nu)$ with range $L^2(B_r)$, in other words the multiplication operator by the characteristic function χ_r of B_r. We recall that $\chi_r(\underline{x}) = 1$ if $\underline{x}\in B_r$ and $\chi_r(\underline{x}) = 0$ if $\underline{x}\notin B_r$.

We notice that

$$\underset{r\to\infty}{\text{s-lim}}\, F_r = I. \tag{5.2}$$

Indeed, for each $f\in L^2(\mathbb{R}^\nu)$,

$$\|(I - F_r)f\|^2 = \int_{|\underline{x}|\ge r} |f(\underline{x})|^2 d^\nu x \to 0 \quad \text{as} \quad r \to \infty.$$

The physical interpretation of F_r is as follows. We assume that the Hilbert space $L^2(\mathbb{R}^\nu)$ is used to describe a (structureless) physical system in ν-dimensional configuration space according to the rules of quantum mechanics [JA]. Then F_r is the observable of localization in B_r; in other words, if $f\in H$, $f \ne \theta$, then

$\|F_r f\|^2/\|f\|^2$ is the probability of finding the system in the state f localized inside the ball B_r. (Each vector $f \neq \theta$ in $\mathcal{H}$ determines a pure state of the system, the unit ray $\{\alpha f | \alpha \in \mathbb{C}, |\alpha| = \|f\|^{-1}\}$. With some abuse of terminology, we shall simply refer to the vectors f themselves as "states".)

Now let $\{U_t\}$ be an evolution group. We denote its infinitesimal generator by H, since we now have in mind the group giving the actual time evolution of a quantum-mechanical system, so that H is the Hamiltonian of the system. A vector f in $\mathcal{H}$ is called a bound state of H if, in the course of its evolution, it remains essentially localized in a bounded region of configuration space at *all* times. Mathematically this requirement may be formulated as follows : given any $\varepsilon > 0$, there is a finite ball B_r such that, for all $t \in \mathbb{R}$, the probability that $U_t f$ be localized outside B_r is less than ε :

$$\frac{1}{\|f\|^2} \int_{|\underline{x}| \geq r} |(U_t f)(\underline{x})|^2 d^\nu x < \varepsilon \qquad \forall\, t \in \mathbb{R}, \quad \text{some} \quad r \in (0,\infty).$$

Written differently, f is a bound state if

$$\lim_{r\to\infty} \sup_{t\in\mathbb{R}} \|(I - F_r)U_t f\|^2 = 0. \tag{5.3}$$

The set of all bound states of H will be denoted by $\mathcal{M}_o(H)$.

One may also introduce two sets $\mathcal{M}_o^+(H)$ and $\mathcal{M}_o^-(H)$ consisting of those states that are bound at positive or negative times respectively :

$$f \in \mathcal{M}_o^\pm(H) \iff \lim_{r\to\infty} \sup_{t\in[0,\pm\infty)} \|(I - F_r)U_t f\|^2 = 0. \tag{5.4}$$

Clearly

$$M_0(H) = M_0^+(H) \cap M_0^-(H). \tag{5.5}$$

We now turn to the definition of the scattering states. In a scattering situation, the system should be localized far away from each bounded region at large (positive and negative) times; it should propagate towards infinity as $t \to \pm\infty$. A priori there is no reason why these scattering states at $t = +\infty$ should be the same those at $t = -\infty$; we shall therefore distinguish between the two cases.

The preceding remarks lead to the following definition : A vector f in H is a scattering state at $t = \pm\infty$ if, for each $r < \infty$,

$$\frac{1}{\|f\|^2} \int_{|\underline{x}|\leq r} |(U_t f)(\underline{x})|^2 d^\nu x \to 0 \quad \text{as} \quad t \to \pm\infty \quad \text{resp.}$$

An equivalent definition is

$$\lim_{t\to\pm\infty} \|F_r U_t f\|^2 = 0 \qquad \forall\, r < \infty. \tag{5.6}$$

The set of all scattering states at $t = \pm\infty$ will be denoted by $M_\infty^\pm(H)$.

We saw in Chapter 4 that evolution groups automatically have certain ergodic properties. In order to make use of these ergodic properties, it is convenient to introduce a somewhat weaker notion of scattering states by doing an average over the parameter t. Instead of requiring the convergence to zero of the probability of finding the system in any finite ball B_r, one simply requires the convergence to zero of the mean squared probability of finding the system in any such ball. Thus we introduce two sets $\overline{M}_\infty^\pm(H)$ of "scattering states on the time average" as follows :

$$f \in \overline{M}_\infty^\pm(H) \iff \lim_{T\to\infty} \frac{1}{T} \cdot \int_0^{\pm T} \|F_r U_t f\|^2 dt = 0 \quad \forall\, r < \infty. \tag{5.7}$$

Remark 5.1 : The use of characteristic functions to define the scattering states in (5.6) and (5.7) is inessential. We made this choice because the operators F_r have a simple physical interpretation. An equivalent definition of $M_\infty^\pm(H)$ would for instance be that

$$\lim_{t\to\pm\infty} \|\varphi(Q)U_t f\|^2 = 0 \qquad \forall \varphi \in C_0^\infty(\mathbb{R}^\nu).$$

In the first proposition below we give some simple properties of the subsets $M_0^\pm(H)$, $M_\infty^\pm(H)$ and $\overline{M}_\infty^\pm(H)$ of the Hilbert space H. In particular we shall see that the vectors in $H_p(H)$ are always bound states, whereas the scattering states are always contained in the subspace of continuity $H_c(H)$ of the Hamiltonian H. The following question then naturally arises : Under what conditions does one have $H_p(H) = M_0(H)$ and $H_c(H) = \overline{M}_\infty^\pm(H)$? We shall give sufficient conditions for these identities to hold and list a few examples.

Proposition 5.2 : (a) $M_0(H)$, $M_0^\pm(H)$, $M_\infty^\pm(H)$ and $\overline{M}_\infty^\pm(H)$ are subspaces of H. Each of these subspaces is invariant under U_t.

(b) $M_\infty^+(H) \subseteq \overline{M}_\infty^+(H)$ and $M_\infty^-(H) \subseteq \overline{M}_\infty^-(H)$.

(c) $M_0^+(H) \perp \overline{M}_\infty^+(H)$ and $M_0^-(H) \perp \overline{M}_\infty^-(H)$.

(d) $M_0(H) \perp \overline{M}_\infty^\pm(H)$.

(e) $H_p(H) \subseteq M_0(H)$ and $\overline{M}_\infty^\pm(H) \subseteq H_c(H)$.

(f) $M_\infty^\pm(H) \subseteq H_w(H)$.

Proof : (b) follows immediately from (5.6), (5.7) and Lemma 4.18 (a), whereas (d) is a consequence of (c) and (5.5).

(a) From (1.12) we obtain

$$\|(I - F_r)U_t(f + \alpha g)\|^2 \le 2\|(I - F_r)U_t f\|^2 + 2|\alpha|^2\|(I - F_r)U_t g\|^2.$$

Hence, if $f, g \in M_0$ and $\alpha \in \mathbb{C}$, then

$$\limsup_{r\to\infty} \sup_{t\in\mathbb{R}} \|(I - F_r)U_t(f + \alpha g)\|^2 = 0,$$

so that $f + \alpha g \in M_0$. Thus M_0 is a linear manifold.

Now let $\{f_n\} \in M_0$ be such that s-lim $f_n = f$. To prove that M_0 is a subspace, we must show that $f \in M_0$. For this, we notice that

$$\|(I - F_r)U_t f\|^2 \leq 2\|(I - F_r)U_t(f - f_n)\|^2 + 2\|(I - F_r)U_t f_n\|^2$$

$$\leq 2\|f - f_n\|^2 + 2\|(I - F_r)U_t f_n\|^2.$$

Given $\varepsilon > 0$, first choose n such that $\|f - f_n\|^2 < \varepsilon/4$. Since $f_n \in M_0$, the second term is less than $\varepsilon/2$ for all $t \in \mathbb{R}$ provided that $r \geq r_0$. Hence

$$\limsup_{r\to\infty} \sup_{t\in\mathbb{R}} \|(I - F_r)U_t f\|^2 = 0.$$

Similarly one shows that $M_0^{\pm}$, $M_\infty^{\pm}$ and $\overline{M}_\infty^{\pm}$ are subspaces. The invariance of these subspaces under U_t follows easily from the definitions (5.3), (5.4), (5.6) and (5.7) (see also (5.17)).

(c) Let $f \in M_0^+$, $g \in \overline{M}_\infty^+$, $\varepsilon > 0$. Then $(f,g) = (U_t f, U_t g)$, hence, by (1.12) and (1.10),

$$|(f,g)|^2 = \frac{1}{T}\int_0^T |(U_t f, U_t g)|^2 dt$$

$$= \frac{1}{T}\int_0^T |(U_t f, F_r U_t g) + ((I - F_r)U_t f, U_t g)|^2 dt$$

$$\leq \frac{2}{T}\|f\|^2 \int_0^T \|F_r U_t g\|^2 dt + \frac{2}{T}\|g\|^2 \int_0^T \|(I - F_r)U_t f\|^2 dt. \tag{5.8}$$

Since $f \in M_0^+$, we may choose $r < \infty$ such that $\|(I - F_r)U_t f\|^2 < \varepsilon(4\|g\|^2)^{-1}$ for all $t \geq 0$. Then the second term on the r.h.s. is less than $\varepsilon/2$ for all $T > 0$. Since $g \in \overline{M}_\infty^+$, this first term converges to zero as $T \to \infty$. Hence it is less than $\varepsilon/2$ for some $T > 0$. Thus $|(f,g)|^2 < \varepsilon$ for each $\varepsilon > 0$, which shows that $f \perp g$.

(e) If f is an eigenvector of H, say $Hf = \lambda f$, then $U_t f = \exp(-i\lambda t)f$, hence $\|(I - F_r)U_t f\|^2 = \|(I - F_r)f\|^2 \to 0$ as $r \to \infty$, by (5.2). Hence each eigenvector of H is in $M_o(H)$. Since $H_p(H)$ is the subspace spanned by all eigenvectors of H and $M_o(H)$ is also a subspace, we have $H_p(H) \subseteq M_o(H)$.

This inclusion, together with (d), now implies that

$$\overline{M}_\infty^\pm(H) \subseteq M_o(H)^\perp \subseteq H_p(H)^\perp = H_c(H). \tag{5.9}$$

(f) Let $f \in M_\infty^+(H)$. Then

$$|(f,U_t f)| = |((I - F_r)f,U_t f) + (f,F_r U_t f)|$$

$$\leq \|(I - F_r)f\| \|f\| + \|f\| \|F_r U_t f\|.$$

Given $\varepsilon > 0$, first choose r such that $\|(I - F_r)f\| < \varepsilon(2\|f\|)^{-1}$ and then t_o such that $\|F_r U_t f\| < \varepsilon(2\|f\|)^{-1}$ for $t \geq t_o$. Thus, given $\varepsilon > 0$, there is a t_o such that $|(f,U_t f)| < \varepsilon$ whenever $t \geq t_o$. This means that $f \in H_w^+(H) = H_w(H)$. □

Proposition 5.3 : Each of the following two conditions implies that $M_o(H) = H_p(H)$, $\overline{M}_\infty^\pm(H) = H_c(H)$ and $M_\infty^\pm(H) = H_w(H)$:

(α) $F_r(H - zI)^{-1} \in B_\infty$ for each $r < \infty$ and some $z \in \rho(H)$,

(β) there is an operator C in $B(H)$ such that $CU_t = U_t C$ for all $t \in \mathbb{R}$, $R(C)$ is dense in H and $F_r C \in B_\infty$ for each $r < \infty$.

Proof : Let $f \in H_c(H)$. By Proposition 4.25, each of the conditions (α) and (β) implies that

$$\lim_{T\to\pm\infty} \frac{1}{T} \int_0^{\pm T} \|F_r U_t f\|^2 dt = 0 \qquad \forall r < \infty.$$

Hence $\mathcal{H}_c(H) \subseteq \overline{\mathcal{M}}_\infty^\pm(H)$. Hence all inclusions in (5.9) must be identities, which proves the first two assertions. Similarly, (4.84) implies that $\mathcal{H}_w(H) \subseteq \mathcal{M}_\infty^\pm(H)$, and the third assertion follows upon combining this with Proposition 5.2 (f). □

Example 5.4 : Let $\varphi: \mathbb{R}^\nu \to \mathbb{R}$ and $H = \varphi(\underline{Q})$ be the multiplication operator in $L^2(\mathbb{R}^\nu)$ by $\varphi(\underline{x})$. Then

$$(U_t f)(\underline{x}) = e^{-i\varphi(\underline{x})t} f(\underline{x}),$$

hence $\|(I - F_r)U_t f\|^2 = \|(I - F_r)f\|^2 \to 0$ as $r \to \infty$. We thus see that, in this example, $\mathcal{M}_o(H) = \mathcal{H}$ and $\overline{\mathcal{M}}_\infty^\pm(H) = \{\theta\}$. In particular, if the spectrum of H is not pure point spectrum, $\mathcal{M}_o(H) \cap \mathcal{H}_c(H) \neq \{\theta\}$. For example, if $H = \underline{Q}^2$: $\mathcal{M}_o(H) = \mathcal{H}_c(H) = \mathcal{H}$.

This example shows that the conclusions of Proposition 5.3 will hold only under suitable hypotheses on H but not in general.

Example 5.5 : Let $\varphi: \mathbb{R}^\nu \to \mathbb{R}$ and $H = \varphi(\underline{P})$ be the multiplication operator in $\tilde{L}^2(\mathbb{R}^\nu)$ by $\varphi(\underline{k})$. Here

$$(FU_t f)(\underline{k}) = e^{-i\varphi(\underline{k})t}\tilde{f}(\underline{k}).$$

Let $C = (|\underline{P}|^{2\nu} + I)^{-1}$ be the multiplication operator by $(|\underline{k}|^{2\nu} + 1)^{-1}$ in $\tilde{L}^2(\mathbb{R}^\nu)$. C commutes with U_t, its range is $\mathcal{R}(C) = D(|\underline{P}|^{2\nu})$, which is dense in $\mathcal{H}$, and $F_r C$ is compact (in fact $F_r C = \chi_r(\underline{Q})(|\underline{P}|^{2\nu} + I)^{-1}$ is Hilbert-Schmidt by Proposition 3.6). Hence, by Proposition 5.3, $\mathcal{M}_o(H) = \mathcal{H}_p(H)$, $\overline{\mathcal{M}}_\infty^\pm(H) = \mathcal{H}_c(H)$ and $\mathcal{M}_\infty^\pm(H) = \mathcal{H}_w(H)$.

Example 5.6 : A special case of Example 5.5 is the Schrödinger free Hamiltonian $H_o = \underline{P}^2$. In this case $\mathcal{M}_o(H_o) = \mathcal{H}_p(H_o) = \{\theta\}$, cf. (2.29). Furthermore, all vectors in $\mathcal{H}$ are scattering states in the sense of (5.6) (without averaging over the parameter t) :

$$\mathcal{M}_\infty^\pm(H_o) = \mathcal{H}_{ac}(H_o) = \mathcal{H}. \qquad (5.10)$$

In fact, we have $\mathcal{M}_\infty^\pm(H_0) = \mathcal{H}_w(H_0)$ by Example 5.5 and $\mathcal{H}_w(H_0) =$ $= \mathcal{H}_{ac}(H_0) = \mathcal{H} = L^2(\mathbb{R}^\nu)$ by Proposition 4.35.

Lemma 5.7 : Let H_0 be given by (2.27). Then, for each $r < \infty$ and each $z \in \rho(H_0)$, $F_r(H_0 - z)^{-1} \in \mathcal{B}_\infty$.

Proof : This is a special case of Corollary 3.14 (b). □

Proposition 5.8 : In $L^2(\mathbb{R}^\nu)$, let $H = H_0 + V$, where $v \in \mathcal{V}_\emptyset$. Then $\mathcal{M}_0(H) = \mathcal{H}_p(H)$, $\overline{\mathcal{M}}_\infty^\pm(H) = \mathcal{H}_c(H)$ and $\mathcal{M}_\infty^\pm(H) = \mathcal{H}_w(H)$.

Proof : By Propositions 2.28 and 2.24, we have

$$F_r(H - i)^{-1} = F_r(H_0 - i)^{-1} - F_r(H_0 - i)^{-1} V(H - i)^{-1},$$

with $V(H - i)^{-1} \in \mathcal{B}(\mathcal{H})$. Since $F_r(H_0 - i)^{-1} \in \mathcal{B}_\infty$, this implies that $F_r(H - i)^{-1} \in \mathcal{B}_\infty$, and the result follows from Proposition 5.3. □

We shall show later (Proposition 5.34) for a subclass of the potentials considered in Proposition 5.8 that $\overline{\mathcal{M}}_\infty^\pm(H) = \mathcal{M}_\infty^\pm(H)$. Another point worth noticing is that the compactness of $F_r(H - i)^{-1}$ used to prove Proposition 5.8 should be true by imposing only local conditions on v (i.e. v could be singular at infinity). This is indeed easy to verify by using the results of Section 2.5 :

Proposition 5.9 : Let $v \in L^p_{loc}(\mathbb{R}^\nu)$ for some p satisfying $p \geq 2$, $p > \nu/2$ (i.e. v satisfies (2.53) for each $\varphi \in C_0^\infty(\mathbb{R}^\nu)$). Let H be an arbitrary self-adjoint extension of the minimal operator $\hat{H} =$ $= -\Delta + v(\underline{x})$ defined on $D(\hat{H}) = C_0^\infty(\mathbb{R}^\nu)$. Then $\mathcal{M}_0(H) = \mathcal{H}_p(H)$, $\overline{\mathcal{M}}_\infty^\pm(H) =$ $= \mathcal{H}_c(H)$ and $\mathcal{M}_\infty^\pm(H) = \mathcal{H}_w(H)$.

Proof : Let $r < \infty$. Choose a function φ in $C_0^\infty(\mathbb{R}^\nu)$ such that $\varphi(\underline{x}) = 1$ for $\underline{x} \in B_r$. Then

$$F_r(H + i)^{-1} = F_r\varphi(\underline{Q})(H + i)^{-1} = F_r(H_0 + i)^{-1}(H_0 + i)\varphi(\underline{Q})(H + i)^{-1}. \quad (5.11)$$

Now $F_r(H_0+i)^{-1}\in\mathcal{B}_\infty$ by Lemma 5.7, whereas $(H_0+i)\varphi(\underline{Q})(H+i)^{-1}\in$ $\in\mathcal{B}(L^2(\mathbb{R}^\nu))$ by Proposition 2.34. Hence $F_r(H+i)^{-1}\in\mathcal{B}_\infty$ by Proposition 3.8 (b), and the result of the present proposition follows from Proposition 5.3. □

Let us now admit, as we did in Section 2.5, that the potential v may also have strong *local* singularities. More precisely, assume that $v\varphi\in L^p_{loc}(\mathbb{R}^\nu)$ ($p\geq 2$, $p>\nu/2$) for all $\varphi\in C_0^\infty(\mathbb{R}^\nu\backslash\Gamma)$, where Γ is a bounded closed set of Lebesgue measure zero. Let H be an arbitrary self-adjoint extension of the minimal operator (2.54), and let Δ be a compact subset of $\mathbb{R}^\nu\backslash\Gamma$ and $F_\Delta=\chi_\Delta(\underline{Q})$ the multiplication operator by the characteristic function χ_Δ of Δ. By choosing a $\varphi\in C_0^\infty(\mathbb{R}^\nu\backslash\Gamma)$ such that $\varphi(\underline{x})=1$ for $\underline{x}\in\Delta$, one sees as in (5.11) that $F_\Delta(H+i)^{-1}\in\mathcal{B}_\infty$. Consequently, by Proposition 4.25,

$$\lim_{t\to\pm\infty}||F_\Delta U_t f||^2=0 \qquad \forall\, f\in\mathcal{H}_w(H), \tag{5.12}$$

$$\lim_{T\to\infty}\frac{1}{T}\int_0^{\pm T}||F_\Delta U_t f||^2dt=0 \qquad \forall\, f\in\mathcal{H}_c(H). \tag{5.13}$$

In other words, if for instance $f\in\mathcal{H}_w(H)$, then the probability of finding the system in any given compact subset of $\mathbb{R}^\nu\backslash\Gamma$ tends to zero. Intuitively there are now two possibilities : either the system propagates away to infinity, i.e. f is a scattering state, or it moves closer and closer to the set Γ, i.e. the state f gets "absorbed" at the singularities of v. This leads to the introduction of four further subsets of $\mathcal{H}$, the sets $\mathcal{M}_\Gamma^\pm(H)$ of <u>absorbed states</u> at $t=\pm\infty$, respectively, and the sets $\overline{\mathcal{M}}_\Gamma^\pm(H)$ of states that are absorbed on the time average. Thus

$$f\in\mathcal{M}_\Gamma^\pm(H) \iff \lim_{t\to\pm\infty}||\varphi(\underline{Q})U_t f||^2=0 \qquad \forall\,\varphi\in C_\Gamma, \tag{5.14}$$

$$f\in\overline{\mathcal{M}}_\Gamma^\pm(H) \iff \lim_{T\to\infty}\frac{1}{T}\int_0^{\pm T}||\varphi(\underline{Q})U_t f||^2dt=0 \qquad \forall\,\varphi\in C_\Gamma. \tag{5.15}$$

The class of functions C_Γ was introduced in Definition 2.36. If Γ is bounded, C_Γ contains all infinitely differentiable functions φ such that $\varphi(\underline{x}) = 1$ near infinity and on any given compact subset Δ_0 of $\mathbb{R}^\nu \setminus \Gamma$, and $\varphi(\underline{x}) = 0$ in some neighbourhood of Γ. Thus, if $f \in M_\Gamma^+(H)$ and Δ is any open neighbourhood of Γ, the probability that $U_t f$ be localized in Δ converges to 1 as $t \to +\infty$. If for example $\Gamma = \{\underline{0}\}$, one has for any $f \in M_\Gamma^+(H)$:

$$\lim_{t\to+\infty} \|F_r U_t f\|^2 / \|f\|^2 = 1 \qquad \forall\, r > 0. \tag{5.16}$$

Since such a state will be essentially localized in a very small region at large times, it is clear from the quantum-mechanical uncertainty relations that it must acquire infinite momentum as $t \to +\infty$. Before proving this, we collect a few simple properties of the subsets $M_\Gamma^\pm$ and $\overline{M}_\Gamma^\pm$ of $\mathcal{H}$.

Proposition 5.10 : Let Γ be a bounded closed set of measure zero. Then

(a) $M_\Gamma^\pm(H)$ and $\overline{M}_\Gamma^\pm(H)$ are subspaces of $\mathcal{H}$ and invariant under U_t.

(b) $M_\Gamma^+(H) \subseteq \overline{M}_\Gamma^+(H)$ and $M_\Gamma^-(H) \subseteq \overline{M}_\Gamma^-(H)$.

(c) $\overline{M}_\infty^+(H) \perp \overline{M}_\Gamma^+(H)$ and $\overline{M}_\infty^-(H) \perp \overline{M}_\Gamma^-(H)$.

(d) $\overline{M}_\Gamma^\pm(H) \subseteq \mathcal{H}_c(H)$.

(e) $M_\Gamma^\pm(H) \subseteq \mathcal{H}_w(H)$.

(f) $M_\Gamma^+(H) \subseteq M_0^+(H)$, $M_\Gamma^-(H) \subseteq M_0^-(H)$.

(g) $M_\Gamma^+(H) \cap M_\Gamma^-(H) \subseteq M_0(H)$.

Proof : The proof of (a), (b), (c) and (e) is very similar to the proof of the corresponding statements in Proposition 5.2 and will be omitted. For (e), one replaces the projections $\{F_r\}$ in the proof of Proposition 5.2 by a sequence $\{\varphi_r\} \in C_\Gamma$ such that $\varphi_r(\underline{x}) \to 1$

$\forall \underline{x} \notin \Gamma$, which implies that s-lim $\varphi_r(\underline{Q}) = I$ as $r \to \infty$. (g) follows from (f) and (5.5).

(d) Let g be an eigenvector of H, say $Hg = \lambda g$, and $f \in \bar{M}_\Gamma^+(H)$. We must show that $(g,f) = 0$. For this, let $\varepsilon > 0$. By the above remark, we may choose a function $\varphi_r \in C_\Gamma$ such that $||[I - \varphi_r(\underline{Q})]g||^2 < \varepsilon(4||f||^2)^{-1}$. Then, as in (5.8)

$$|(f,g)|^2 \le \frac{2}{T}||f||^2 \int_0^T ||[I - \varphi_r(\underline{Q})]U_t g||^2 dt + \frac{2}{T}||g||^2 \int_0^T ||\varphi_r(\underline{Q})U_t f||^2 dt.$$

Since $||[I - \varphi_r(\underline{Q})]U_t g|| = ||[I - \varphi_r(\underline{Q})]g||$, the first term is less than $\varepsilon/2$ for each $T > 0$. The second term converges to zero as $T \to \infty$, hence it is less than $\varepsilon/2$ for some T. Thus $|(f,g)|^2 < \varepsilon$ for each $\varepsilon > 0$, i.e. $(f,g) = 0$.

(f) For $f \in H$, set $h_r(t) = ||(I - F_r)U_t f||^2$. Each h_r is a continuous function of t, and the sequence $\{h_r(t)\}$ decreases monotonically to zero for each fixed $t \in \mathbb{R}$. By Dini's Theorem [R], the sequence $\{h_r\}$ converges uniformly to zero on any closed finite interval Δ : Given $\varepsilon > 0$, there is a number $r_o = r_o(\varepsilon, \Delta)$ such that

$$\sup_{t \in \Delta} ||(I - F_r)U_t f||^2 < \varepsilon \qquad \forall\, r \ge r_o. \tag{5.17}$$

Now let $f \in M_\Gamma^+(H)$. Choose a ball B_R containing Γ in its interior. Then $||(I - F_R)U_t f|| \to 0$ as $t \to +\infty$. Hence there is a number T such that $||(I - F_r)U_t f||^2 < \varepsilon$ for all $r \ge R$ and all $t \ge T$. Taking $\Delta = [0,T]$ in (5.17), we see that $f \in M_o^+$. □

We now show that, if $f \in M_\Gamma^\pm$, then it acquires infinite momentum, hence infinite kinetic energy, as $t \to \pm\infty$ respectively, and if $f \in \bar{M}_\Gamma^\pm$, it acquires infinite momentum on the time average (the kinetic energy operator is just the free Hamiltonian $H_o = \underline{P}^2$).

Proposition 5.11 : Let Γ be a bounded closed set of measure zero, $v \in L_{loc}^p(\mathbb{R}^\nu \setminus \Gamma)$ with $p \ge 2$, $p > \nu/2$, and H an arbitrary self-adjoint extension of the minimal operator $\hat{H}$.

(a) If $f_\pm \in \overline{M}^\pm_\Gamma(H)$, then for each $\psi \in C_0^\infty(\mathbb{R})$:

$$\lim_{T\to\infty} \frac{1}{T} \int_0^{\pm T} \|\psi(H_0) U_t f_\pm\|^2 dt = 0. \tag{5.18}$$

(b) If $g_\pm \in M^\pm_\Gamma(H)$, then, for each $\psi \in C_0^\infty(\mathbb{R})$:

$$\text{s-}\lim_{t\to\pm\infty} \psi(H_0) U_t g_\pm = \theta. \tag{5.19}$$

Remark : In $\tilde{L}^2(\mathbb{R}^\nu)$, (5.19) implies that, for each $M < \infty$:

$$\lim_{t\to\pm\infty} \int_{|\underline{k}|\le M} |(FU_t g_\pm)(\underline{k})|^2 d^\nu k = 0.$$

This means that, as $t \to \pm\infty$, the probability that the momentum or the kinetic energy of the vector $U_t g_\pm$ stays in any finite set tends to zero; in other words $U_t g_\pm$ propagates to infinity in momentum space.

Proof : (a) Let B_R be a ball containing Γ in its interior. Let $\varphi \in C_0^\infty(\mathbb{R}^\nu)$ be such that $\varphi(\underline{x}) = 0$ for $\underline{x} \in B_R$ and $\varphi(\underline{x}) = 1$ for $|\underline{x}| \ge 2R$. Then

$$[I - \bar{\varphi}(Q)]\bar{\psi}(H_0) = [I - \bar{\varphi}(Q)](H_0 + I)^{-1}(H_0 + I)\bar{\psi}(H_0),$$

which is in B_∞ by Corollary 3.14 (b) and Proposition 4.11. Consequently $\psi(H_0)[I - \varphi(Q)] \in B_\infty$. Now by (1.12)

$$\frac{1}{T}\left|\int_0^{\pm T} \|\psi(H_0)U_t f_\pm\|^2 dt\right| \le \frac{2}{T}\left|\int_0^{\pm T} \|\psi(H_0)[I - \varphi(Q)]U_t f_\pm\|^2 dt\right|$$

$$+ \|\psi\|_\infty^2 \frac{2}{T}\left|\int_0^{\pm T} \|\varphi(Q) U_t f_\pm\|^2 dt\right|.$$

As $T \to \infty$, the first term on the r.h.s. converges to zero by Proposition 4.25 and the second one because $f_\pm \in \overline{M}^\pm_\Gamma(H)$.

(b) One has, with φ as above :

$$\psi(H_o)U_t g_\pm = \psi(H_o)[I - \varphi(\underline{Q})]U_t g_\pm + \psi(H_o)\varphi(\underline{Q})U_t g_\pm .$$

Since $U_t g_\pm$ converges weakly to zero as $t \to \pm\infty$ by Proposition 5.10 (e), the first term on the r.h.s. converges strongly to zero by Proposition 3.9. The second term converges strongly to zero as $t \to \pm\infty$ because $g_\pm \in M_\Gamma^\pm(H)$. □

We end this section with a few more remarks about local absorption. Since H commutes with U_t and formally $H = \underline{P}^2 + V$, we expect that absorption should be possible only if the potential v has a sufficiently large negative part near Γ; indeed, one has formally

$$(f,Hf) = (U_t f, HU_t f) = (U_t f, H_o U_t f) + (U_t f, VU_t f). \tag{5.20}$$

If $f \in M_\Gamma^+(H)$, the first term on the r.h.s. tends to $+\infty$ as $t \to +\infty$ by Proposition 5.11, the second one to $(U_t f, VF_\Delta U_t f)$, where Δ is any open neighbourhood of Γ. Since the l.h.s. is independent of t, one should have $(U_t f, VF_\Delta U_t f) \to -\infty$ as $t \to +\infty$.

The preceding argument has only heuristic value though. In fact we must choose $f \in D(H)$ in (5.20); this implies that $U_t f \in D(H)$ for each t, but in general $U_t f$ will be neither in $D(H_o)$ nor in D(V). It has been pointed out in [4] that, by suitably interpreting Pearson's example ([5], see below), one may produce local absorption even with $v(\underline{x}) \equiv 0$ by taking for Γ a countable union of spheres and defining H by suitable boundary conditions on these spheres.

On the other hand a potential that simply converges very rapidly to $-\infty$ as $\underline{x}$ approaches Γ will usually not lead to absorption either. (This has been proved rigorously only in special cases, for example for spherically symmetric extensions of $\hat{H}$ for potentials of the form $v(\underline{x}) = \alpha|\underline{x}|^{-n}$, $\alpha < 0$ and n arbitrary, see [6,7].) The intuition behind this is that, as a particle approaches Γ, there will be a very strong, purely attractive force exerted on it, and it will just shoot through the singularity at infinite speed.

The situation is different if v is more and more rapidly oscillating as $\underline{x}$ approaches Γ, with divergent amplitude for the oscillations. A particle approaching Γ is then under the influence of both strong attractive and strong repulsive forces varying rapidly as a function of $\underline{x}$. It may then happen that it gets trapped in this oscillating field for an infinite length of time, which is the phenomenon of absorption. A potential of this type, for which $\dim M_\Gamma^{\pm}(H) = \infty$, was explicitly constructed by Pearson [5]. The oscillations of v near Γ have to be carefully chosen : it has been shown for a class of oscillating potentials that no absorption occurs [8,9].

A final comment concerns the time average in defining scattering states and absorbed states. The averaging over time was introduced in order to allow us to use the Mean Ergodic Theorem and its consequences in proving properties of the subsets of scattering states and absorbed states. The time average is related to spectral properties of H and is needed only to treat vectors in $H_{sc}(H)$ which do not belong to $H_w(H)$. In many cases no such vectors exist. In fact, we shall show further on in this chapter, for a class of Schrödinger operators H, that $M_o(H) = H_p(H)$ and $M_\infty^{\pm}(H) = H_c(H) = H_{ac}(H)$.

5.2 Wave Operators

We have seen that the scattering states associated with a Hamiltonian H propagate to infinity as the time t tends to $+\infty$ or $-\infty$. Often the potential is rapidly tending to zero at large distances, so that a scattering state should feel only a very small force when $|t|$ is very large. Thus one expects that at large times a scattering state should behave almost like a free state. The first goal of scattering theory is to give a more detailed description of the time evolution of scattering states at large times by showing that they can be better and better approximated by the time evolution of a different state under a different and simpler evolution group (called the *free evolution group*). If both of these evolution groups are given, the possibility of such an asymptotic approximation of one of them by the other one is not a priori guaranteed; it is a condition on the interaction or the potential (formally the difference between the two infinitesimal generators), called the asymptotic condition.

From now on we shall therefore consider two evolution groups $\{U_t\}$ and $\{U_t^o\}$, called total evolution group and free evolution group respectively. For the moment we let $\{U_t^o\}$ be an abstract group, but further on in these lectures $\{U_t^o\}$ will be the Schrödinger free evolution group introduced in Chapter 4.4. The infinitesimal generators of $\{U_t\}$ and $\{U_t^o\}$ will be denoted by H and H_o and called the total Hamiltonian and the free Hamiltonian respectively. Thus, in the sense of Proposition 4.4 :

$$U_t = \exp(-iHt), \qquad U_t^o = \exp(iH_o t). \tag{5.21}$$

We begin by formulating the asymptotic condition in mathematical terms. As already said, if $f \in \mathcal{M}_\infty^+(H)$ is a scattering state of H (at $t = +\infty$), there should exist a vector $g_+ \in \mathcal{M}_\infty^+(H_o)$ such that

$$\lim_{t\to+\infty} \|U_t f - U_t^o g_+\| = 0. \tag{5.22}$$

Since U_t is unitary, one has

$$\|U_t f - U_t^o g_+\| = \|U_t^*(U_t f - U_t^o g_+)\| = \|f - U_t^* U_t^o g_+\|. \tag{5.23}$$

Thus (5.22) may be rewritten as

$$\lim_{t\to+\infty} \|f - U_t^* U_t^o g_+\| = 0, \tag{5.24}$$

i.e. $$f = \operatorname{s-lim}_{t\to+\infty} U_t^* U_t^o g_+. \tag{5.25}$$

In the same way, if $f \in \mathcal{M}_\infty^-(H)$, there should exist a vector $g_- \in \mathcal{M}_\infty^-(H_o)$ such that

$$f = \operatorname{s-lim}_{t\to-\infty} U_t^* U_t^o g_-. \tag{5.26}$$

These considerations lead naturally to the introduction of the following two operators $\Omega_\pm$, called the <u>wave operators</u> :

$$\Omega_\pm = \operatorname{s-lim}_{t\to\pm\infty} U_t^* U_t^o E_\infty^\pm(H_o), \tag{5.27}$$

where $E_\infty^\pm(H_o)$ denotes the orthogonal projection with range $\mathcal{M}_\infty^\pm(H_o)$. In the present section we assume that the limits in (5.27) exist and derive some general properties of $\Omega_\pm$. Conditions for the existence of these limits will be indicated in the following two sections.

For the general properties of $\Omega_\pm$ given below, it is inessential that the limits of $U_t^* U_t^o$ be considered on the subspaces $\mathcal{M}_\infty^\pm(H_o)$. We shall therefore replace $E_\infty^\pm(H_o)$ in (5.27) by a general projection $E_\pm$ commuting with U_t^o, i.e. we assume throughout this section that

$$E_\pm = E_\pm^* = E_\pm^2, \qquad U_t^o E_\pm = E_\pm U_t^o \qquad \forall t \in \mathbb{R}, \tag{5.28}$$

and that the following limits exists :

$$\Omega_\pm = \operatorname{s-lim}_{t\to\pm\infty} U_t^* U_t^o E_\pm. \tag{5.29}$$

Proposition 5.12 : $\Omega_\pm$ are partial isometries satisfying

$$\Omega_\pm^*\Omega_\pm = E_\pm . \tag{5.30}$$

If $E_\pm = I$, then $\Omega_\pm$ are isometries.

Proof : (i) We assumed that the limits (5.29) exist on each $f \in \mathcal{H}$. Thus $D(\Omega_\pm) = \mathcal{H}$.

(ii) Let $h \in \mathcal{H}$. One has as in (5.23) that

$$\lim_{t\to+\infty} \| U_t^{o*} U_t \Omega_+ h - E_+ h \| = \lim_{t\to+\infty} \| \Omega_+ h - U_t^* U_t^o E_+ h \| = 0 ,$$

hence
$$\operatorname*{s-lim}_{t\to+\infty} U_t^{o*} U_t \Omega_+ = E_+ . \tag{5.31}$$

Now let $f,g \in \mathcal{H}$. Then, by (5.28), (5.31) and Proposition 1.1 :

$$(f,E_+ g) = (E_+ f, E_+ g) = \lim_{t\to+\infty} (E_+ f, U_t^{o*} U_t \Omega_+ g)$$

$$= \lim_{t\to+\infty} (U_t^* U_t^o E_+ f, \Omega_+ g) = (\Omega_+ f, \Omega_+ g) = (f, \Omega_+^* \Omega_+ g).$$

Thus, for each fixed g, $E_+ g - \Omega_+^*\Omega_+ g$ is orthogonal to all $f \in \mathcal{H}$, so that $E_+ g - \Omega_+^*\Omega_+ g = \theta$ by Lemma 1.5. This shows that $\Omega_+^*\Omega_+ = E_+$. Similarly one gets $\Omega_-^*\Omega_- = E_-$. □

Proposition 5.13 : $\Omega_\pm$ intertwine H and H_o :

(a) $$U_t \Omega_\pm = \Omega_\pm U_t^o \qquad \forall\, t \in \mathbb{R} , \tag{5.32}$$

$$\Omega_\pm^* U_t = U_t^o \Omega_\pm^* \qquad \forall\, t \in \mathbb{R}. \tag{5.33}$$

(b) If φ is such that $\tilde{\varphi} \in L^1(\mathbb{R})$, then

$$\varphi(H)\Omega_\pm = \Omega_\pm \varphi(H_o), \tag{5.34}$$

$$\Omega_\pm^* \varphi(H) = \varphi(H_o)\Omega_\pm^*. \tag{5.35}$$

(c) If $f \in D(H_o)$, then $\Omega_\pm f \in D(H)$ and

$$H\Omega_\pm f = \Omega_\pm H_o f. \tag{5.36}$$

Proof : (a) We use (4.5) and (4.3) and set $\tau = s - t$:

$$U_t \Omega_+ = U_t \cdot \underset{s\to+\infty}{\text{s-lim}}\, U_s^* U_s^o E_+ = \underset{s\to+\infty}{\text{s-lim}}\, U_t U_s^* U_s^o E_+$$

$$= \underset{s\to+\infty}{\text{s-lim}}\, U_{s-t}^* U_s^o E_+ = \underset{\tau\to+\infty}{\text{s-lim}}\, U_\tau^* U_\tau^o U_t^o E_+ = \Omega_+ U_t^o ,$$

which proves (5.32). Upon taking its adjoint, one gets $\Omega_\pm^* U_t^* = U_t^{o*}\Omega_\pm^*$ for each $t \in \mathbb{R}$, and (5.33) follows by setting $t = -t'$ and noticing that $U_{-t'}^* = U_{t'}$.

(b) follows immediately from (a) and (4.37). For (c), we use (a) and Proposition 4.1 to obtain that

$$\frac{i}{\tau}(U_\tau - I)\Omega_\pm f = \frac{i}{\tau}\Omega_\pm (U_\tau^o - I)f \;\to\; \Omega_\pm H_o f \quad \text{as} \quad \tau \to 0.$$

Hence $\{i\tau^{-1}(U_\tau - I)\Omega_\pm f\}$ is strongly convergent as $\tau \to 0$, so that $\Omega_\pm f \in D(H)$ by Proposition 4.1. Again by Proposition 4.1, this strong limit is $H\Omega_\pm f$, which proves (5.36). □

Let us now define

$$F_\pm = \Omega_\pm \Omega_\pm^*. \tag{5.37}$$

By Proposition 1.20, $F_\pm$ are the projections onto the ranges $\mathcal{R}(\Omega_\pm)$ of Ω_+ and Ω_- respectively. The subspaces $\mathcal{R}(\Omega_\pm) = \mathcal{R}(F_\pm)$ are invariant under the total evolution group; this is the content of part (a) of the following proposition :

Proposition 5.14 :

(a) $U_t F_\pm = F_\pm U_t \quad \forall\, t\in\mathbb{R}.$ (5.38)

(b) If φ is such that $\tilde{\varphi}\in L^1(\mathbb{R})$, then

$$\varphi(H)F_\pm = F_\pm\varphi(H). \tag{5.39}$$

In particular :

$$g\in R(\Omega_\pm) \;\Rightarrow\; \varphi(H)g\in R(\Omega_\pm), \tag{5.40}$$

$$g\perp R(\Omega_\pm) \;\Rightarrow\; \varphi(H)g\perp R(\Omega_\pm). \tag{5.41}$$

(c) If $R(E_\pm)\subseteq H_c(H_o)$, then $R(F_\pm)\subseteq H_c(H)$.

(d) If $R(E_\pm)\subseteq H_{ac}(H_o)$, then $R(F_\pm)\subseteq H_{ac}(H)$.

(e) If $R(E_\pm)\subseteq H_w(H_o)$, then $R(F_\pm)\subseteq H_w(H)$.

(f) If $R(E_\pm)\subseteq M_\infty^\pm(H_o)$, then $R(F_\pm)\subseteq M_\infty^\pm(H)$.

Proof : (a) This follows easily from (5.32) and (5.33) :

$$U_tF_\pm = U_t\Omega_\pm\Omega_\pm^* = \Omega_\pm U_t^o\Omega_\pm^* = \Omega_\pm\Omega_\pm^* U_t = F_\pm U_t.$$

(b) (5.39) is an immediate consequence of (a) and (4.37). Furthermore, if $F_+g = g$, then by (5.39) : $F_+\varphi(H)g = \varphi(H)F_+g = \varphi(H)g$, which proves (5.40). On the other hand, if $g\perp R(\Omega_+)$, then $(I - F_+)g = g$, so that again by (5.39) : $(I - F_+)\varphi(H)g = \varphi(H)(I - F_+)g = \varphi(H)g$, which shows that $\varphi(H)g\perp R(\Omega_+)$.

(c) Let g be an eigenvector of H, say $Hg = \lambda g$. Let $f\in R(E_\pm)$. Then, for each $t\in\mathbb{R}$:

$$|(g,\Omega_+f)| = |(U_tg,U_t\Omega_+f)| = |e^{i\lambda t}(g,\Omega_+U_t^of)| = |(\Omega_+^*g,U_t^of)|. \tag{5.42}$$

Since $f\in H_c(H_o)$, we obtain from (5.42) and Proposition 4.23 that

$$|(g,\Omega_+f)| = \lim_{T\to\infty} \frac{1}{T}\int_0^T |(g,\Omega_+f)|dt = \lim_{T\to\infty}\frac{1}{T}\int_0^T |(\Omega_+^*g,U_t^0 f)|dt = 0.$$

Hence $R(\Omega_\pm)$ is orthogonal to each eigenvector g of H; thus $R(\Omega_\pm)\subseteq H_c(H)$.

(d,e) Let $f\in R(E_+)$ and $g = \Omega_+f\in R(F_+)$. Then, by (5.32) and (5.30) :

$$(g,U_tg) = (\Omega_+f,U_t\Omega_+f) = (\Omega_+^*\Omega_+f,U_t^0f) = (f,U_t^0f).$$

Thus, if $(f,U_t^0f) \to 0$ or $(f,U_t^0f)\in L^2(\mathbb{R})$, then $(g,U_tg) \to 0$ or $(g,U_tg)\in L^2(\mathbb{R})$ respectively.

(f) Let $h\in R(\Omega_+)$, i.e. $h = \text{s-lim}\, U_t^*U_t^0f$ as $t \to +\infty$, for some $f\in R(E_+)\subseteq M_\infty^+(H_0)$. Then, by (1.12),

$$||F_rU_th||^2 \le 2||F_r||^2||U_th - U_t^0f||^2 + 2||F_rU_t^0f||^2.$$

We have $||F_r|| = 1$. Given any $\varepsilon > 0$, there is a number T such that, for all $t \ge T$:

(α) $||U_th - U_t^0f||^2 < \varepsilon/4$,

(β) $||F_rU_t^0f||^2 < \varepsilon/4$ (since $f\in M_\infty^+(H_0)$).

Hence $||F_rU_th||^2 < \varepsilon$ for all $t \ge T$, i.e. $h\in M_\infty^+(H)$. □

Corollary 5.15 : Let H_0 be the Schrödinger free Hamiltonian (2.27). Then, for any $E_\pm$,

$$R(\Omega_\pm)\subseteq M_\infty^\pm(H) \cap H_{ac}(H). \qquad (5.43)$$

Proof : By (5.10) we have $R(E_\pm)\subseteq H = H_{ac}(H_0) = M_\infty^\pm(H_0)$, so that (5.43) holds by Proposition 5.14 (d) and (f). □

Proposition 5.14 and Corollary 5.15 lead naturally to the notion of asymptotic completeness. One may consider various forms of this notion, some of which we shall now briefly discuss.

We assume that, as is usually the case, one has $R(E_\pm) \subseteq H_c(H_0)$. Then $R(\Omega_\pm) \subseteq H_c(H)$ by Proposition 5.14 (c). One then calls the wave operators $\Omega_\pm$ complete (and speaks of asymptotic completeness of the theory) if the preceding inclusions are both equalities, i.e. if

$$R(\Omega_+) = R(\Omega_-) = H_c(H) = H_p(H)^\perp. \tag{5.44}$$

In this case one has a complete asymptotic description, as $t \to \pm\infty$, of all vectors in $H_c(H)$ in terms of the free evolution group U_t^0, in the sense of (5.22) and its analogue for $t \to -\infty$; in order words, if $f \in H_c(H)$, there are two vectors $g_\pm \in R(E_\pm)$ such that

$$\lim_{t\to\pm\infty} \| U_t f - U_t^0 g_\pm \| = 0. \tag{5.45}$$

In other situations the ranges of $\Omega_\pm$ will be only subspaces of $H_c(H)$. One expects that quite generally $R(\Omega_\pm) = M_\infty^\pm(H)$ or $R(\Omega_\pm) = \overline{M}_\infty^\pm(H)$. One may still speak of asymptotic completeness, in a generalized sense, provided that one can give a description of the behaviour of the vectors in $H_c(H) \cap R(\Omega_\pm)^\perp$ as $t \to \pm\infty$. A simple situation is that where all vectors in $H_c(H) \cap R(\Omega_\pm)^\perp$ are bound states, i.e. where $H_p(H)$ in (5.44) is simply replaced by $M_0(H)$:

$$R(\Omega_+) = R(\Omega_-) = M_0(H)^\perp. \tag{5.46}$$

This is more general than (5.44), since $H_p(H)$ may be a proper subspace of $M_0(H)$, and has been termed asymptotic completeness in the geometric sense in [10].

A somewhat weaker notion is obtained by considering (5.46) for each sign of time separately, i.e. by requiring that

$$H = R(\Omega_+) \oplus M_0^+(H) = R(\Omega_-) \oplus M_0^-(H), \tag{5.47}$$

which may be called asymptotic completeness in the geometric sense at positive and at negative times. Here $R(\Omega_+)$ may be different from $R(\Omega_-)$. This type of asymptotic completeness implies ordinary asymptotic completeness in the geometric sense if and only if the bound states at positive times and at negative times are the same, i.e. if and only if $M_o^+(H) = M_o^-(H)$.

The term "generalized asymptotic completeness" is used if all states in $H_c(H) \cap R(\Omega_\pm)^\perp$ get absorbed as $t \to \pm\infty$ [11]. Thus we speak of generalized asymptotic completeness if

$$H_c(H) = R(\Omega_+) \oplus M_\Gamma^+(H) = R(\Omega_-) \oplus M_\Gamma^-(H) \tag{5.48}$$

and of generalized asymptotic completeness on the average if

$$H_c(H) = R(\Omega_+) \oplus \overline{M}_\Gamma^+(H) = R(\Omega_-) \oplus \overline{M}_\Gamma^-(H). \tag{5.49}$$

Here again the ranges of Ω_+ and Ω_- will in general not be the same. Notice that, by Proposition 5.10 (f), generalized asymptotic completeness implies asymptotic completeness in the geometric sense at positive and at negative times (at least if $R(E_\pm) \subseteq \overline{M}_\infty^\pm(H_o)$).

It is sometimes useful to consider a more general type of wave operators in which there appears an additional factor between the two evolution groups. Let $J_\pm$ be two operators in $B(H)$. Then one may consider the limits

$$W_\pm = \underset{t\to\pm\infty}{\text{s-lim}}\ U_t^* J_\pm U_t^o E_\pm. \tag{5.50}$$

We use the symbol $W_\pm$ for these more general wave operators and $\Omega_\pm$ for the case where $J_\pm = I$. An example that will be discussed in Section 5.4 is that where $J_\pm$ are "cut-off operators", cutting off a part of configuration space. Another situation where such operators $J_\pm$ occur is that where U_t^o and U_t act in two different

Hilbert spaces, say $\mathcal{H}_0$ and $\mathcal{H}$ respectively. $J_\pm$ are then some "injection operators" mapping $\mathcal{H}_0$ into $\mathcal{H}$.

We end this section by giving some properties of $W_\pm$.

Proposition 5.16 : Assume that the limits in (5.50), defining $W_\pm$, exist. Then

(a) $W_\pm$ intertwine H and H_0 :

$$U_t W_\pm = W_\pm U_t^0 , \qquad W_\pm^* U_t = U_t^0 W_\pm^* \qquad \forall t \in \mathbb{R} \tag{5.51}$$

and, if $\tilde{\varphi} \in L^1(\mathbb{R})$:

$$\varphi(H) W_\pm = W_\pm \varphi(H_0), \qquad W_\pm^* \varphi(H) = \varphi(H_0) W_\pm^* . \tag{5.52}$$

(b) If $\mathcal{R}(E_\pm) \subseteq \mathcal{H}_c(H_0)$, then $\mathcal{R}(W_\pm) \subseteq \mathcal{H}_c(H)$.

(c) If $\mathcal{R}(E_\pm) \subseteq \mathcal{H}_w(H_0)$, then $\mathcal{R}(W_\pm) \subseteq \mathcal{H}_w(H)$.

(d) If s-lim $F_r J_\pm U_t^0 E_\pm = 0$ as $t \to \pm\infty$ $\forall r < \infty$, then $\mathcal{R}(W_\pm) \subseteq \mathcal{M}_\infty^\pm(H)$.

(e) The range of $W_\pm$, its orthogonal complement $\mathcal{R}(W_\pm)^\perp$ and its closure are invariant under U_t. (Contrarily to $\mathcal{R}(\Omega_\pm)$, $\mathcal{R}(W_\pm)$ need not be closed.)

Proof : The proof of (a)-(d) is identical with that of the corresponding statements in Propositions 5.13 and 5.14 (for (c), use also Proposition 4.24 (b)). To prove (e), let $g \in \mathcal{R}(W_+)$, i.e. $g = W_+ f$ for some $f \in \mathcal{R}(E_+)$. Then, by (5.51), $U_t g = W_+ U_t^0 f$, hence $U_t g \in \mathcal{R}(W_+)$. Thus $\mathcal{R}(W_+)$ is invariant under U_t. Similarly $U_t^* g = U_{-t} g = W_+ U_{-t}^0 f \in \mathcal{R}(W_+)$, and the remaining two statements follow from Lemma 1.18. □

As a final result we establish a relation between $W_\pm$ and $\Omega_\pm$:

Proposition 5.17 : Assume that

$$\operatorname*{s-lim}_{t\to\pm\infty}(J_\pm - I)U_t^oE_\pm = 0. \tag{5.53}$$

Then

(a) If $W_\pm$ exists, then $\Omega_\pm$ also exists and $W_\pm = \Omega_\pm$. In particular $W_\pm$ are partial isometries with $W_\pm^*W_\pm = E_\pm$.

(b) If $\Omega_\pm$ exists, then so does $W_\pm$, and $W_\pm = \Omega_\pm$.

Proof : (a) We have by the triangle inequality and the unitarity of U_t^* that

$$||W_\pm f - U_t^*U_t^oE_\pm f|| \le ||W_\pm f - U_t^*J_\pm U_t^oE_\pm f|| + ||(J_\pm - I)U_t^oE_\pm f||,$$

which converges to zero as $t \to \pm\infty$ since $W_\pm$ exist. Hence $\{U_t^*U_t^oE_\pm f\}$ is strongly convergent to $W_\pm f$, i.e. $\Omega_\pm$ exist and $\Omega_\pm = W_\pm$. The proof of (b) is similar. □

5.3 Abstract Conditions for Existence and Completeness of Wave Operators

The purpose of this section is to prove an abstract theorem on existence and completeness of wave operators. In the following section we shall then verify the conditions of this theorem for Schrödinger operators.

Throughout this section we assume that $J_\pm$ are two operators in $\mathcal{B}(\mathcal{H})$ and $E_\pm$ two projections satisfying (5.28). We first derive existence criteria for the limits $W_\pm = \text{s-lim } U_t^*J_\pm U_t^oE_\pm$ as $t \to \pm\infty$ (Propositions 5.19 and 5.20). From Lemma 1.8 (applied to the situation where the discrete index n is replaced by the continuous index t), we see that it is enough to prove the existence of s-lim $U_t^*J_\pm U_t^oE_\pm$ on some total subset $\mathcal{D}_o^\pm$ of $\mathcal{R}(E_\pm)$. The next lemma gives a sufficient condition for s-lim f(t) as $t \to \infty$ to exist.

Lemma 5.18 : Let $\{W_t\}_{t\geq 0}$ be a family of bounded operators. Suppose $\{W_t f\}$ is strongly continuously differentiable and that the following real-valued improper Riemann integral exists for some $a < \infty$:

$$\int_a^\infty \|\frac{d}{dt}W_t f\| dt < \infty.$$

Then $\{W_t f\}$ is strongly convergent as $t \to +\infty$. A similar result holds for $t \to -\infty$.

Proof : By (1.61) and (1.57), we have for $t \geq \tau$:

$$\|W_t f - W_\tau f\| = \|\int_\tau^t \frac{d}{ds}W_s f\, ds\| \leq \int_\tau^t \|\frac{d}{ds}W_s f\|\, ds.$$

As $\tau, t \to \infty$, the last integral converges to zero by hypothesis. Hence $\{W_t f\}$ is strongly Cauchy. □

Proposition 5.19 : Suppose $J_\pm$ map $D(H_o)$ into $D(H)$ and that $t \mapsto \|(HJ_\pm - J_\pm H_o)U_t^o f\|$ is integrable at $t = \pm\infty$ for all f in some subset $\mathcal{D}_o^\pm$ of $\mathcal{R}(E_\pm)$ which is total in $\mathcal{R}(E_\pm)$ and contained in $D(H_o)$. Then the wave operators $W_\pm = \text{s-}\lim_{t\to\pm\infty} U_t^* J_\pm U_t^o E_\pm$ exist.

Proof : By using Proposition 4.1, Corollary 4.3 and the rule for differentiating a product (which, as is easily checked, may be applied in the present situation), we get for $g \in D(H_o)$:

$$\frac{d}{dt}U_t^* J_\pm U_t^o g = U_t^* iHJ_\pm U_t^o g + U_t^* J_\pm(-iH_o)U_t^o g = iU_t^*(HJ_\pm - J_\pm H_o)U_t^o g. \quad (5.54)$$

Hence $\|\frac{d}{dt}U_t^* J_\pm U_t^o g\| \leq \|(HJ_\pm - J_\pm H_o)U_t^o g\|$.

If $g \in \mathcal{D}_o^\pm$, this is integrable at $t = \pm\infty$, so that $U_t^* J_\pm U_t^o$ is strongly convergent on $\mathcal{D}_o^\pm$ as $t \to \pm\infty$ by Lemma 5.18. Since $\mathcal{D}_o^\pm$ is total in $\mathcal{R}(E_\pm)$, Lemma 1.8 implies the existence of $W_\pm$ on $\mathcal{R}(E_\pm)$. (Remark : the strong continuity of the derivative in (5.54) is not essential for this argument but can also be checked.) □

For the next result we introduce the notations $R_z = (H-z)^{-1}$, $R_z^o = (H_o - z)^{-1}$ and

$$Y_z^{\pm} = J_{\pm}R_z^o - R_z J_{\pm}. \tag{5.55}$$

From the identity $A(A-z)^{-1} = I + z(A-z)^{-1}$, one easily deduces that

$$Y_z^{\pm} = HR_z J_{\pm} R_z^o - R_z J_{\pm} H_o R_z^o = R_z(HJ_{\pm} - J_{\pm}H_o)R_z^o. \tag{5.56}$$

The first equality holds on all of $\mathcal{H}$ (since HR_z and $H_oR_z^o$ are defined everywhere), whereas the second one is true only under the additional assumption that $J_{\pm}D(H_o) \subseteq D(H)$.

Proposition 5.20 : Suppose there is a total subset $\mathcal{D}_o^{\pm}$ of $\mathcal{R}(E_{\pm})$ such that, for each $f \in \mathcal{D}_o^{\pm}$ and some $z \in \rho(H) \cap \rho(H_o)$:

$$\left|\int_0^{\pm\infty} ||Y_z^{\pm} U_f^o f|| dt\right| < \infty. \tag{5.57}$$

Then $W_{\pm} = \text{s-}\lim_{t\to\pm\infty} U_t^* J_{\pm} U_t^o E_{\pm}$ exists.

Proof : We give the proof for $t \to +\infty$ and omit the superscript or subscript $+$ on $\mathcal{D}_o^+, Y_z^+, J_+$ and E_+.

(i) By using (1.15) we see that

$$\left| ||Y_z U_{t+\tau}^o f|| - ||Y_z U_t^o f|| \right| \le ||Y_z(U_{t+\tau}^o - U_t^o)f|| \le ||Y_z|| \, ||(U_\tau^o - I)f||.$$

This implies that the function $t \mapsto ||Y_z U_t^o f||$ is uniformly continuous on $\mathbb{R}$, for each $f \in \mathcal{H}$. For $f \in \mathcal{D}_o$, this function is also integrable at $+\infty$ by the hypothesis (5.57). It then follows from Lemma A.5 that $||Y_z U_t^o f|| \to 0$ as $t \to +\infty$, in other terms that $\text{s-}\lim Y_z U_t^o f = \theta$ as $t \to +\infty$ $\forall f \in \mathcal{D}_o$. Since $\mathcal{D}_o$ is dense in $\mathcal{R}(E)$ and $||Y_z U_t^o|| \le ||Y_z|| < \infty$, we obtain by applying Lemma 1.8 that $\text{s-}\lim Y_z U_t^o f = \theta$ $\forall f \in \mathcal{R}(E)$, i.e. that

$$\text{s-lim}_{t\to+\infty} Y_z U_t^o E = 0. \tag{5.58}$$

(5.58) implies that $U_t^* Y_z U_t^o E \to 0$ strongly as $t \to +\infty$. After insertion of the definition (5.55) of Y_z, this reads

$$\text{s-lim}_{t\to\infty}(U_t^* J U_t^o R_z^o E - R_z U_t^* J U_t^o E) = 0. \tag{5.59}$$

(ii) By a calculation as in (5.54) and by using (5.56), we obtain for any $f \in H$:

$$\frac{d}{dt} R_z U_t^* J U_t^o R_z^o f = \frac{d}{dt} U_t^* R_z J U_t^o R_z^o f$$

$$= iU_t^* H R_z J U_t^o R_z^o f - iU_t^* R_z J U_t^o H_o R_z^o f = iU_t^* Y_z U_t^o f. \tag{5.60}$$

Thus, if $f \in D_o$,

$$\int_0^\infty \left\| \frac{d}{dt} R_z U_t^* J U_t^o R_z^o f \right\| dt = \int_0^\infty \| Y_z U_t^o f \| dt < \infty.$$

Hence, by Lemma 5.18, $R_z U_t^* J U_t^o R_z^o f$ is strongly convergent as $t \to \infty$, for each $f \in D_o$. Since $\| R_z U_t^* J U_t^o R_z^o \| \leq \| R_z J R_z^o \| < \infty$, s-lim $R_z U_t^* J U_t^o R_z^o f$ exists for each $f \in R(E)$ by Lemma 1.8. Now $R_z^o R(E) = D(A) \cap R(E)$ is dense in $R(E)$ by Lemma 4.7 (c), so that Lemma 1.8 implies the strong convergence of $R_z U_t^* J U_t^o E$ as $t \to \infty$. By combining this with (5.59), we see that $U_t^* J U_t^o R_z^o E$ is strongly convergent. Using again, as above, the denseness of $R_z^o R(E)$ in $R(E)$, one obtains the strong convergence of $U_t^* J U_t^o$ on $R(E)$, which proves the existence of $W_+ = \text{s-lim}\, U_t^* J U_t^o E$ as $t \to +\infty$. □

Proposition 5.19 shows that, in order for the wave operators $W_\pm$ to exist, it suffices to know that the function $t \mapsto \| (HJ_\pm - J_\pm H_o) U_t^o f \|$ is integrable at $t = \pm\infty$ for sufficiently many vectors f. In applications, U_t^o is often explicitly given in some representation of H as a function space (consider e.g. the Schrödinger free evolution group), and so is $HJ_\pm - J_\pm H_o$. It is then rather easy to verify the above-mentioned integrability condition.

Similarly Proposition 5.20 requires the integrability at $t = \pm\infty$ of the function $t \mapsto \|Y_z U_t^o f\|$ for sufficiently many vectors f. In what follows, we make a stronger integrability assumption, essentially that $t \mapsto \|Y_z^{\pm} U_t^o E_{\pm} D_{\pm}\|$ be integrable at $\pm\infty$, where $D_{\pm}$ are two suitably chosen operators (essentially, $D_{\pm}$ must form a partition of unity, i.e. $D_+ + D_- = I$, and satisfy some further conditions). While this assumption is too strong if one just wants to prove the existence of the wave operators, it is suitable for establishing simultaneously the asymptotic completeness of the theory and various spectral properties of the total Hamiltonian H.

We will now state a set of conditions which imply existence and completeness of the wave operators. We shall then, in a sequence of lemmas, draw various conclusions from some of these conditions and finally sum up, in Proposition 5.28, the general results implied by the entire set of conditions.

For the sake of simplicity, we assume from now on that $E_+ = E_-$. Thus, in the conditions (C1)-(C8) below, E is a projection commuting with U_t^o and such that $R(E) \subseteq \mathcal{H}_c(H_o)$; $D_{\pm}, J_{\pm}$ and $K_{\pm}$ are operators in $\mathcal{B}(\mathcal{H})$ and z is some complex number in $\rho(H) \cap \rho(H_o)$. The conditions involving a function φ are assumed to hold for each $\varphi \in C_{oo}^{\infty}(\mathbb{R})$.

(C1) $E = E(D_+ + D_-)$.

(C2) a) $\underset{t\to\pm\infty}{\text{s-lim}}\, D_{\mp} U_t^o E = 0$,

b) $\underset{t\to\pm\infty}{\text{s-lim}}\, D_{\mp}^* U_t^o E = 0$.

(C3) $\left| \int_0^{\pm\infty} \|Y_z^{\pm} \varphi(H_o) U_t^o E D_{\pm}\| dt \right| < \infty$.

(C4) $Y_Z^{\pm} \in \mathcal{B}_\infty$.

(C5) $(J_+ - J_-)R_Z^o E \in \mathcal{B}_\infty$.

(C6) $J_\pm \varphi(H_o)(I - E) \in \mathcal{B}_\infty$.

(C7) $\lim_{T\to\infty} \frac{1}{T} \int_0^{\pm T} ||K_\pm(I - J_\pm^*)U_t f||^2 dt = 0 \qquad \forall\, f \in \mathcal{H}_c(H)$.

(C8) $\text{s-}\lim_{t\to\pm\infty} (J_\pm - I)U_t^o E = 0$.

Let us make a few more comments on these conditions. (C1) essentially means that D_+^* and D_-^* form a partition of unity on the subspace $\mathcal{R}(E)$. (C2) is most easily interpreted when $D_\pm$ are projections. It then means that, under the free evolution group, each vector in $\mathcal{R}(E)$ eventually has no component left in the range of D_- as $t \to +\infty$, i.e. it propagates out of the subspace $\mathcal{R}(D_-)$, and similarly for D_+ and $t \to -\infty$. (C3) is the integrability condition already mentioned. (C4) means that the difference of H and H_o must be small in a certain sense, cf. (5.56). (C5) requires that J_+ and J_- are not too different. It trivially holds when $J_+ = J_-$. Similarly (C6) is trivial when E = I. It means that E should not be too different from I and implies a restriction on the point spectrum of H_o which we shall discuss later. (C8) states that $J_\pm$ hardly changes a vector in $\mathcal{R}(E)$ that has evolved for a very long time under the free evolution group. (C8) becomes trivial when $J_\pm = I$.

All the conditions that we have mentioned so far involve, apart from E and $D_\pm$, only the free Hamiltonian and the difference of H and H_o. (C7) is different in this respect, as it is a hypothesis on the *total* evolution group. It is a condition of evanescence, as considered in Section 5.1. If for instance $K_\pm = I$ and $J_\pm = I - F_r$, it means that all vectors in $\mathcal{H}_c(H)$ are evanescent, on the time average, from the ball B_r (in the framework of Section 5.1). If r may take all values in $(0,\infty)$, then (C7), with $K_\pm = I$, is the requirement that $\mathcal{H}_c(H) = \overline{\mathcal{M}}^{\pm}_{\infty}(H)$. If $J_\pm = I$, then (C7) becomes trivial.

<u>Lemma 5.21</u> : (a) Assume (C1), (C2,a) and (C3). Then the wave operators $W_\pm = \text{s-lim } U_t^* J_\pm U_t^o E$ as $t \to \pm\infty$ exist.

(b) Assume in addition (C8). Then $\Omega_\pm = \text{s-lim } U_t^* U_t^o E$ as $t \to \pm\infty$ also exist, and $W_\pm = \Omega_\pm$.

<u>Proof</u> (for the + sign) : (a) Let $g \in \mathcal{R}(E)$, $s \geq 0$ and $\varphi \in C_{00}^{\infty}(\mathbb{R})$. Then (setting $\tau = t - s$) :

$$\int_0^\infty \| Y_z U_t^o E\varphi(H_o) U_s^{o*} D_+ U_s^o g \| dt \leq \|g\| \int_{-s}^\infty \| Y_z U_\tau^o E\varphi(H_o) D_+ \| d\tau .$$

The last integral is finite by (C3) and since $\|Y_z U_\tau^o E\varphi(H_o) D_+\| \leq$ $\leq \|Y_z\| \|\tilde{\varphi}\|_1 \|D_+\|$, which is integrable on $[-s,0]$. Hence, by Proposition 5.20, W_+ exists on the subspace $\mathcal{M}_+$ of $\mathcal{R}(E)$ spanned by $\{E\varphi(H_o) U_s^{o*} D_+ U_s^o g \,|\, \varphi \in C_{00}^\infty(\mathbb{R}),\ s \geq 0,\ g \in \mathcal{R}(E)\}$.

It now suffices to show that $\mathcal{M}_+ = \mathcal{R}(E)$. Since $\varphi(H_o)E =$ $= E\varphi(H_o)$, it suffices to prove, by virtue of Proposition 4.27, that $\mathcal{N}_+ \equiv \{EU_s^{o*} D_+ U_s^o g \,|\, s \geq 0,\ g \in \mathcal{R}(E)\}$ is total in $\mathcal{R}(E)$. So let $f \in \mathcal{R}(E)$. Then, by (C1), $f = EU_s^{o*}(D_+ + D_-) U_s^o f$. This implies that

$$\| f - EU_s^{o*} D_+ U_s^o f \| = \| EU_s^{o*} D_- U_s^o f \| \leq \| D_- U_s^o f \| ,$$

which converges to zero as $s \to +\infty$ by (C2,a). This shows that $\mathcal{N}_+$ is total (in fact dense) in $\mathcal{R}(E)$.

(b) This was shown in Proposition 5.17. □

<u>Lemma 5.22</u> : Assume that (C4) holds. Then one has for each φ such that φ and $\tilde{\varphi}$ are in $L^1(\mathbb{R})$:

$$\varphi(H)J_{\pm} - J_{\pm}\varphi(H_o)\in\mathcal{B}_{\infty} \tag{5.61}$$

and $$(J_{\pm} - U_t^* J_{\pm} U_t^o)\varphi(H_o)\in\mathcal{B}_{\infty}. \tag{5.62}$$

Proof : (i) One has the following identity, which follows easily from the definition (5.55) of $Y_z^{\pm}$: if $\zeta\in\rho(H)\cap\rho(H_o)$, then

$$Y_{\zeta}^{\pm} = (H-z)R_{\zeta}Y_z^{\pm}(H_o - z)R_{\zeta}^o.$$

Since $(H-z)R_{\zeta} = I + (\zeta - z)R_{\zeta}\in\mathcal{B}(\mathcal{H})$ and similarly $(H_o - z)R_{\zeta}^o\in\mathcal{B}(\mathcal{H})$, and since $Y_z^{\pm}\in\mathcal{B}_{\infty}$ by hypothesis, we have $Y_{\zeta}^{\pm}\in\mathcal{B}_{\infty}$ for each $\zeta\in\rho(H)\cap\rho(H_o)$.

(ii) To prove (5.61), we use Lemma 4.12. From the expression (4.50) for $\varphi_{\varepsilon}(A)$ and the definition of $Y_z^{\pm}$, one finds that

$$\varphi_{\varepsilon}(H)J_{\pm} - J_{\pm}\varphi_{\varepsilon}(H_o) = (2\pi i)^{-1}\int_{-\infty}^{\infty}\varphi(\lambda)(Y_{\lambda-i\varepsilon}^{\pm} - Y_{\lambda+i\varepsilon}^{\pm})d\lambda. \tag{5.63}$$

By (2.17), the function $\lambda\mapsto Y_{\lambda\pm i\varepsilon}^{\pm}$ is continuous in the operator norm. Since φ, the inverse Fourier transform of a L^1-function, is also continuous, it follows that the integral in (5.63) exists as a Riemann integral in the operator norm. Since the integrand belongs to $\mathcal{B}_{\infty}$ by (i) and $\mathcal{B}_{\infty}$ is complete in the operator norm, the integral is in $\mathcal{B}_{\infty}$ (cf. Remark 1.25). Hence $\varphi_{\varepsilon}(H)J_{\pm} - J_{\pm}\varphi_{\varepsilon}(H_o)\in\mathcal{B}_{\infty}$ for each $\varepsilon > 0$. As $\varepsilon\to 0$, $\varphi_{\varepsilon}(H)J_{\pm} - J_{\pm}\varphi_{\varepsilon}(H_o)$ converges to $\varphi(H)J_{\pm} - J_{\pm}\varphi(H_o)$ in norm, by (4.51). Hence $\varphi(H)J_{\pm} - J_{\pm}\varphi(H_o)\in\mathcal{B}_{\infty}$ by Proposition 3.8 (d), which proves (5.61).

(iii) Let $\psi_t(\lambda) = e^{-i\lambda t}\varphi(\lambda)$. Then $\tilde{\psi}_t(s) = \tilde{\varphi}(t+s)$, and one obtains from (4.37) that $\psi_t(H) = U_t\varphi(H)$, $\psi_t(H_o) = U_t^o\varphi(H_o)$. If we now apply (5.61) for the function ψ_t, we get that $U_t\varphi(H)J_{\pm} - J_{\pm}U_t^o\varphi(H_o)\in\mathcal{B}_{\infty}$. Upon multiplication by U_t^*, we find that $\varphi(H)J_{\pm} - U_t^*J_{\pm}U_t^o\varphi(H_o)\in\mathcal{B}_{\infty}$, and (5.62) follows by combining this with (5.61). □

Lemma 5.23 : Assume (C1), (C2,a), (C3) and (C4). Then, for each $\varphi\in C_{oo}^{\infty}(\mathbb{R})$:

$$R_z(J_\pm - W_\pm)\varphi(H_o)ED_\pm \in \mathcal{B}_\infty. \qquad (5.64)$$

Proof : By Lemma 5.21, $W_\pm$ exist. Also

$$R_z(J_+ - W_+)\varphi(H_o)ED_+ = \underset{t\to+\infty}{\text{s-lim}}\, R_z(J_+ - U_t^* J_+ U_t^o)\varphi(H_o)ED_+$$

$$:= \underset{t\to+\infty}{\text{s-lim}}\, X_t^+.$$

By Lemma 5.22, $X_t^+ \in \mathcal{B}_\infty$ for each t. To prove that the operator on the l.h.s. is in $\mathcal{B}_\infty$, it therefore suffices to show that the limit exists in the operator norm. For this, we notice that, by (5.60) :

$$||R_z(J_+ - W_+)\varphi(H_o)ED_+ - X_t^+|| = ||\int_t^\infty \frac{d}{ds} R_z U_s^* J_+ U_s^o \varphi(H_o)ED_+ ds||$$

$$\leq \int_t^\infty ||Y_z^+ U_s^o (H_o - z)\varphi(H_o)ED_+||\, ds. \qquad (5.65)$$

Now, if $\varphi \in C_{oo}^\infty(\mathbb{R})$, then $(\lambda - z)\varphi \in C_{oo}^\infty(\mathbb{R})$. Hence the above improper integral is finite by (C3), which proves that each term in (5.65) converges to zero as $t \to +\infty$. □

Lemma 5.24 : Assume (C4) and (C5). Then $R_z(J_+ - J_-)E \in \mathcal{B}_\infty$.

Proof : This follows immediately from the identity

$$R_z(J_+ - J_-)E = Y_z^- E - Y_z^+ E + (J_+ - J_-)R_z^o E. \quad □ \qquad (5.66)$$

Lemma 5.25 : Assume (C1), (C2,a,b) and (C3)-(C6). Suppose f is such that $f \in \mathcal{H}_c(H) \cap D(H)$ and $f \perp \mathcal{R}(W_+)$. Then, for each $\varphi \in C_{oo}^\infty(\mathbb{R})$,

$$\lim_{T\to\infty} \frac{1}{T} \int_0^T ||J_+^* \varphi(H) U_t f||\, dt = 0. \qquad (5.67)$$

Proof : The proof is achieved by writing $J_+^*\varphi(H)U_t f$ as a sum of seven conveniently chosen terms. We set $g = (H - \bar{z})f$, so that, by (4.48), $f = R_{\bar{z}} g = R_z^* g$. Then

$$J_+^*\varphi(H)U_tf = [J_+^*\varphi(H) - \varphi(H_o)J_+^*]U_tf + [I - (D_+^* + D_-^*)E]\varphi(H_o)J_+^*U_tf$$

$$+ D_+^*E\varphi(H_o)(J_+^* - W_+^*)R_z^*U_tg + D_-^*E\varphi(H_o)(J_-^* - W_-^*)R_z^*U_tg$$

$$+ D_-^*\varphi(H_o)E(J_+^* - J_-^*)R_z^*U_tg$$

$$+ D_+^*E\varphi(H_o)U_t^oW_+^*f + D_-^*U_t^oE\varphi(H_o)W_-^*f \ , \qquad (5.68)$$

where we have used the intertwining relation (5.51) in the last two terms.

We must show that the time average on [0,T] of the norm of each term on the r.h.s. of (5.68) converges to zero as $T \to \infty$. For the first five terms, this follows from Proposition 4.25, since $f,g \in \mathcal{H}_c(H)$ and the operators in front of U_tf or U_tg are all in $\mathcal{B}_\infty$ (by (5.61) for the first one, (C1) and (C6) for the second one, (5.64) for the next two and by Lemma 5.24 for the fifth one). The sixth term is identically zero, since the hypothesis $f \perp \mathcal{R}(W_+)$ implies that $W_+^*f = \theta$ (cf. Proposition 1.13). Finally the last term converges to zero in the ordinary sense, hence also on the average, by (C2,b). □

We know from Proposition 5.16 that $\mathcal{R}(W_\pm)$ and $\mathcal{R}(W_\pm)^\perp$ are invariant under U_t and that $\mathcal{R}(W_\pm) \subseteq \mathcal{H}_c(H)$. Therefore $\mathcal{G}_\pm :=$ $:= \mathcal{R}(W_\pm)^\perp \cap \mathcal{H}_c(H)$ are subspaces of $\mathcal{H}_c(H)$ which are invariant under U_t, and the problem of asymptotic completeness or generalized asymptotic completeness consists in studying these subspaces $\mathcal{G}_\pm$. In particular the wave operators $W_\pm$ are complete if and only if $\mathcal{G}_\pm = \{\theta\}$, i.e. $\mathcal{R}(W_\pm)^\perp = \mathcal{H}_p(H)$.

We now derive some properties of the subspaces $\mathcal{G}_\pm$.

<u>Lemma 5.26</u> : Assume (C1)-(C6). Then one has for each $f \in \mathcal{G}_\pm$:

$$\lim_{T\to\infty} \frac{1}{T} \int_0^{\pm T} \|J_\pm^* U_t f\|^2 dt = 0. \qquad (5.69)$$

Proof (for the + sign) : Let $f_i \in D(H) \cap G_+$, $\alpha_i \in \mathbb{C}$, $\varphi_i \in C_{00}^\infty(\mathbb{R})$ and $g = \sum_{i=1}^n \alpha_i \varphi_i(H) f_i$ $(n < \infty)$. Then, by the triangle inequality and Lemma 5.25,

$$\lim_{T\to\infty} \frac{1}{T} \int_0^T \|J_+^* U_t g\| dt \le \sum_{i=1}^n |\alpha_i| \lim_{T\to\infty} \frac{1}{T} \int_0^T \|J_+^* \varphi_i(H) U_t f_i\| dt = 0. \quad (5.70)$$

Now $D(H) \cap G_+$ is dense in G_+ by Lemma 4.7 (c) (take B to be the projection with range G_+ in Lemma 4.7). Hence, by Proposition 4.27, the linear span of $\{\varphi(H) D(H) \cap G_+ | \varphi \in C_{00}^\infty(\mathbb{R})\}$ is dense in G_+. Thus (5.70) holds for all g in a dense subset $\mathcal{D}$ of G_+. It is then straightforward to see that

$$\lim_{T\to\infty} \frac{1}{T} \int_0^T \|J_+^* U_t f\| dt = 0$$

for each $f \in G_+$. This, together with Lemma 4.18 (c), implies (5.69).□

Lemma 5.27 : Assume (C1)-(C7). Then one has for each $f \in G_\pm$:

$$\lim_{T\to\infty} \frac{1}{T} \int_0^{\pm T} \|K_\pm U_t f\|^2 dt = 0. \qquad (5.71)$$

Proof : This follows immediately from (C7) and Lemma 5.26 by using the inequality (1.12). □

Proposition 5.28 : (a) Assume (C1)-(C7) with $K_\pm = I$. Then the wave operators $W_\pm$ = s-lim $U_t^* J_\pm U_t^o E$ as $t \to \pm\infty$ exist and are complete in the sense that $\overline{R(W_\pm)} = H_p(H)^\perp$. Furthermore, if $R(E) \subseteq H_w(H_o)$, then $H_c(H) = H_w(H)$.

(b) Assume (C1)-(C8) with $K_\pm = I$. Then the wave operators $\Omega_\pm$ = = s-lim $U_t^* U_t^o E$ as $t \to \pm\infty$ exist and are partial isometries satisfying $\Omega_\pm^* \Omega_\pm = E$ and $\Omega_\pm \Omega_\pm^* = E_c(H)$ (in other words $\Omega_\pm$ are unitary operators from $R(E) \subseteq H_c(H_o)$ onto $H_c(H)$). If $R(E) \subseteq H_{ac}(H_o)$, then

$\mathcal{H}_c(H) = \mathcal{H}_{ac}(H)$, i.e. H has no singularly continuous spectrum.

Proof : Let $f \in G_\pm$. Since $K_\pm = I$, we have by Lemma 5.27 that

$$0 = \lim_{T\to\infty} \frac{1}{T} \int_0^{\pm T} ||U_t f||^2 dt = \pm||f||^2.$$

Hence $f = \theta$, so that $G_\pm = \{\theta\}$. But this means that $\overline{R(W_\pm)} = \mathcal{H}_c(H)$.

If $R(E) \subseteq \mathcal{H}_w(H)$, then $R(W_\pm) \subseteq \mathcal{H}_w(H)$ by Proposition 5.16. Since $\mathcal{H}_w(H) \subseteq \mathcal{H}_c(H)$, we must have $\mathcal{H}_w(H) = \mathcal{H}_c(H)$. (b) follows from (a), Lemma 5.21 (b) and Proposition 5.14 (d). □

We end this section with a result on the point spectrum of H and a few remarks on the conditions (C1)-(C8).

Proposition 5.29 : Assume (C1)-(C6) with $J_+ = I$. Then each non-zero eigenvalue of H is of finite multiplicity (i.e. the corresponding eigensubspace is finite-dimensional), and the only possible accumulation points of the eigenvalues of H are $\lambda = 0$ and $\lambda = \pm\infty$.

Proof : Suppose that the assertion of the proposition is not true. Then there is a real number $\lambda \neq 0$ and an infinite orthonormal sequence $\{e_k\}$ of eigenvectors of H (i.e. $He_k = \lambda_k e_k$) such that $\lambda_k \to \lambda$ as $k \to \infty$. (This is clear from Proposition 2.5 if λ were an accumulation point of the eigenvalues of H; on the other hand, if λ were an eigenvalue of infinite multiplicity, one could take for $\{e_k\}$ an infinite orthonormal basis of the associated eigensubspace, in which case $\lambda_k = \lambda$ for all k.)

We have w-lim $e_k = \theta$ as $k \to \infty$ by Example 1.2. Similarly w-lim$(H - \bar{z})e_k = \theta$, since $(g,(H - \bar{z})e_k) = (\lambda_k - \bar{z})(g,e_k) \to 0$ for each $g \in \mathcal{H}$. We now replace in the identity (5.68) the vector $U_t f$ by e_k and $U_t g$ by $(H - \bar{z})e_k$. The last two terms are then identically zero since $R(W_\pm) \perp H_p(H)$, hence $W^*_\pm e_k = \theta$ by Proposition 1.13. By Proposition 3.9, the first five terms converge strongly to zero as $k \to \infty$ because, as we pointed out in the proof of Lemma 5.25, the operators appearing in these terms are all compact. Thus (since $J_+ = I$), s-lim $\varphi(H)e_k$ = s-lim $\varphi(\lambda_k)e_k = \theta$ for each $\varphi \in C^\infty_{oo}(\mathbb{R})$. If we choose φ such that $\varphi(\mu) = 1$ in a neighbourhood of $\mu = \lambda$, we have $\lim\|e_k\| = 0$ as $k \to \infty$, a contradiction because $\|e_k\| = 1$. □

Remark 5.30 : (a) The conditions (C3) and (C6) were required to hold for all $\varphi \in C^\infty_{oo}(\mathbb{R})$. The choice of this particular class of functions was made simply for the sake of convenience and because it is a suitable class for the application that interests us here, namely Schrödinger operators. In view of Remark 4.28, the proof of asymptotic completeness goes through if one requires (C3) and (C6) to hold only for all $\varphi \in C^\infty_o(\mathbb{R}\backslash\Gamma)$, where Γ is a closed countable set. Under these weaker hypotheses, the conclusion of Proposition 5.29 becomes somewhat different : each eigenvalue $\lambda \notin \Gamma$ of H is of finite multiplicity, and the only possible accumulation points of the eigenvalues of H are the points λ in Γ and $\pm\infty$.

(b) We now comment briefly on the condition (C6), assuming that $J_\pm = I$. This condition then reads $\varphi(H_o)(I - E) \in B_\infty$ for all $\varphi \in C^\infty_{oo}(\mathbb{R})$ or all $\varphi \in C^\infty_o(\mathbb{R}\backslash\Gamma)$. Let us also assume that the projection E is the maximal admissible one, namely $E = E_c(H_o)$. (C6) then becomes the condition that $\varphi(H_o)E_p(H_o) \in B_\infty$ for all φ in one of the classes indicated above. This condition requires that $\varphi(\lambda) = 0$ at each λ which is an eigenvalue of H_o of infinite multiplicity as well as at each λ which is an accumulation point of eigenvalues of H_o. Because Γ must be countable and closed, this shows that H_o cannot be arbitrary (for instance, an operator H_o having a dense set of eigenvalues in some interval Δ is excluded).

(c) It is also possible to show that the hypotheses of Proposition 5.28 (a) together with the additional assumption that $\mathcal{R}(E) \subseteq \mathcal{H}_{ac}(H_o)$, imply that $\mathcal{H}_c(H) = \mathcal{H}_{ac}(H)$, so that H has no singularly continuous spectrum [10]. The condition $\mathcal{R}(E) \subseteq \mathcal{H}_{ac}(H_o)$ is always satisfied when H_o itself has no singularly continuous spectrum, i.e. when $\mathcal{H} = \mathcal{H}_p(H_o) \oplus \mathcal{H}_{ac}(H_o)$. In particular this is true for the Schrödinger free Hamiltonian (2.28), for which one has $\mathcal{H} = \mathcal{H}_{ac}(H_o)$ by Proposition 4.35.

(d) As can be seen from the proof of Lemma 5.25, the compactness assumptions in (C4), (C5) and (C6) are somewhat too strong for proving asymptotic completeness. To obtain the conclusion of Lemma 5.25, it is sufficient to know that each of the operators in front of $U_t f$ or $U_t g$ in the first five terms of (5.68) belongs to $\mathcal{B}_H^{\pm}$, where $\mathcal{B}_H^{\pm}$ is the following subset of $\mathcal{B}(\mathcal{H})$:

$$A \in \mathcal{B}_H^{\pm} \iff \lim_{T\to\infty} \frac{1}{T} \int_0^{\pm T} \|AU_t f\|^2 dt = 0 \qquad \forall f \in \mathcal{H}_c(H).$$

We notice that (α) $\mathcal{B}_H^{\pm}$ are *linear* subsets of $\mathcal{B}(\mathcal{H})$, (β) $\mathcal{B}_H^{\pm}$ are left-invariant under $\mathcal{B}(\mathcal{H})$, i.e. if $A \in \mathcal{B}_H^{\pm}$ and $D \in \mathcal{B}(\mathcal{H})$, then $DA \in \mathcal{B}_H^{\pm}$, ($\gamma$) $\mathcal{B}_H^{\pm}$ are right-invariant under the commutant of H, i.e. if $A \in \mathcal{B}_H^{\pm}$, $D \in \mathcal{B}(\mathcal{H})$ and $DU_t = U_t D$ $\forall t$, then $AD \in \mathcal{B}_H^{\pm}$, ($\delta$) $\mathcal{B}_H^{\pm}$ are closed in the uniform operator topology, i.e. if $A_n \in \mathcal{B}_H^{\pm}$ and $\|A_n - A\| \to 0$, then $A \in \mathcal{B}_H^{\pm}$, ($\varepsilon$) $\mathcal{B}_\infty \subseteq \mathcal{B}_H^{\pm}$.

Suppose now that we replace (C4)-(C6) by the weaker assumptions that $(Y_z^{\pm})^* \in \mathcal{B}_H^+ \cap \mathcal{B}_H^-$, $ER_z^o(J_+^* - J_-^*) \in \mathcal{B}_H^+ \cap \mathcal{B}_H^-$ and $(I - E)\varphi(H_o)J_{\pm}^* \in \mathcal{B}_H^{\pm}$ respectively. By using (α)-(δ) above, one can easily adapt the proofs of Lemmas 5.22-5.24 to deduce that the adjoints of the operators in (5.61), (5.62), (5.64) and (5.66) are in $\mathcal{B}_H^+ \cap \mathcal{B}_H^-$.

5.4 Asymptotic Completeness for Schrödinger Operators

We now apply the abstract results of the preceding section to Schrödinger operators with short range potentials. We first define the class of potentials that we shall consider, next verify the conditions (C1)-(C8), and then state a number of results on asymptotic completeness. In this section we have $\mathcal{H} = L^2(\mathbb{R}^\nu)$.

Definition 5.31 : Let Γ be a bounded closed set of Lebesgue measure zero in $\mathbb{R}^\nu$, and let $\kappa > 0$. Then a measurable function $v : \mathbb{R}^\nu \to \mathbb{R}$ is said to be of class $\mathcal{V}_{\Gamma,\kappa}$ if it can be written in the form $v(\underline{x}) = w(\underline{x})(1+|\underline{x}|)^{-\kappa}$ with $w \in \mathcal{V}_\Gamma$ (see Definition 2.36).

There will be essential differences between the cases $\Gamma = \emptyset$ and $\Gamma \neq \emptyset$ and between the cases $\kappa > 1$ and $\kappa \leq 1$. When $\Gamma = \emptyset$, the local singularities of v (if any) must be in $L^p(\mathbb{R}^\nu)$, whereas in the case $\Gamma \neq \emptyset$ these singularities may be very strong. We shall therefore speak about weakly singular and strongly singular potentials. When $\kappa > 1$, the function $|w_1(\underline{x})|(1+|\underline{x}|)^{-\kappa}$ tends to zero at infinity faster than the Coulomb potential $v_C = \gamma|x|^{-1}$. One then speaks of short range potentials. If $\kappa \leq 1$ and $|w_1(\underline{x})| \geq c > 0$ in a neighbourhood of infinity, the decrease is not more rapid than $|\underline{x}|^{-1}$. Such potentials are called long range potentials.

The free Hamiltonian is $H_0 = \underline{P}^2$, and the total Hamiltonian H is defined as in Chapter 2. All results of Chapter 2 may be applied here since $\mathcal{V}_{\Gamma,\kappa} \subset \mathcal{V}_\Gamma$. When $\Gamma = \emptyset$, $H_0 + V$ is self-adjoint on $D(H_0)$. On the other hand, when $\Gamma \neq \emptyset$, H will be an arbitrary self-adjoint extension of the minimal operator $\hat{H}$ given in (2.54).

We now define the operators E and $J_\pm$. The projection E will be taken to be the identity operator I. This is admissible since H_0 has purely absolutely continuous spectrum, so that $\mathcal{R}(E) = \mathcal{H} = \mathcal{H}_{ac}(H_0)$. With this choice of E, the condition (C6) is trivially satisfied.

When $\Gamma = \emptyset$, we also set $J_\pm = I$. (C7) and (C8) are then trivially satisfied. When $\Gamma \neq \emptyset$, we choose a positive number M such that Γ is contained in the interior of the ball $B_M = \{\underline{x}\in\mathbb{R}^\nu \,|\, |\underline{x}| \le M\}$, and we take $J_+ = J_- = j(\underline{Q})$ to be the multiplication operator by a real function $j\in C^\infty(\mathbb{R}^\nu)$ such that $j(\underline{x}) = 0$ for $|\underline{x}| \le M + 1$ and $j(\underline{x}) = 1$ for $|\underline{x}| \ge M + 2$. In this case it may happen that (C7) is not satisfied, a point that will be discussed later. (C8) however is satisfied also in this case; in fact one has $I - j(\underline{Q}) = [I - j(\underline{Q})]F_{M+2}$, whence for each $f\in\mathcal{H}$:

$$||[I - j(\underline{Q})]U_t^o f|| \le (1 + ||j||_\infty)||F_{M+2}U_t^o f||.$$

This tends to zero as $t \to \pm\infty$ by Example 5.6, which proves (C8).

Notice that the operator $J = j(\underline{Q})$ "cuts off" that part of configuration space where the potential v may be strongly singular. We also remark that, in the case $\Gamma \neq \emptyset$, the cut-off potential $v(\underline{x})j(\underline{x})$ belongs to $\mathcal{V}_{\emptyset,\kappa}$:

$$v\in\mathcal{V}_{\Gamma,\kappa}, \quad \Gamma \neq \emptyset \Rightarrow vj\in\mathcal{V}_{\emptyset,\kappa}. \tag{5.72}$$

The condition (C5) is trivial both for $\Gamma = \emptyset$ and $\Gamma \neq \emptyset$. We next verify (C4). Notice that $Y_z^+ = Y_z^- := Y_z$.

<u>Lemma 5.32</u> : Assume $v\in\mathcal{V}_{\Gamma,\kappa}$. Then $Y_z\in\mathcal{B}_\infty$.

<u>Proof</u> : (i) If $\Gamma = \emptyset$, we have from the expression (5.56) for Y_z :

$$Y_z = R_z V R_z^o = [R_z W][(I + |\underline{Q}|)^{-\kappa} R_z^o]. \tag{5.73}$$

The first factor on the r.h.s. is in $\mathcal{B}(\mathcal{H})$ by Lemma 2.31, whereas the second one is in $\mathcal{B}_\infty$ by Corollary 3.14. Hence their product Y_z is in $\mathcal{B}_\infty$ by Proposition 3.8 (b).

(b) If $\Gamma \neq \emptyset$, we let Φ be the multiplication operator by a function $\varphi(\underline{x})\in C^\infty(\mathbb{R}^\nu)$ such that $\varphi(\underline{x}) = 0$ for $|\underline{x}| \le M$ and $\varphi(\underline{x}) = 1$

for $|\underline{x}| \geq M + 1$. We then have $\varphi(\underline{x})j(\underline{x}) = j(\underline{x})$, in other words $\Phi J = J$. Also $(\Delta\varphi)(\underline{x})j(\underline{x}) = 0$ and $[\partial\varphi(\underline{x})/\partial x_m]j(\underline{x}) = 0$, since $j = 0$ on the support $\{\underline{x} | M \leq |\underline{x}| \leq M + 1\}$ of the derivatives of φ. Thus, in the notation introduced before Proposition 2.20 : $(\Delta\Phi)J = 0$ and $\Phi_{,m}J = 0$.

It follows from this and Proposition 2.20 that for each $f \in D(H_o)$:

$$\Phi H_o Jf - H_o Jf = -(\Delta\Phi)Jf + 2i \sum_{m=1}^{\nu} P_m \Phi_{,m} Jf = \theta. \tag{5.74}$$

Using this, (5.56), Proposition 2.32 and (2.35), we get

$$Y_z = R_z(HJ - JH_o)R_z^o = R_z VJR_z^o + R_z\Phi(H_oJ - JH_o)R_z^o$$

$$= [R_z WJ + R_z\Phi(I + |\underline{Q}|)^\kappa(\Delta J) - 2i \sum_{m=1}^{\nu} R_z\Phi P_m(1 + |\underline{Q}|)^\kappa J_{,m}] \cdot$$

$$\cdot [(I + |\underline{Q}|)^{-\kappa} R_z^o]. \tag{5.75}$$

The second factor is in $\mathcal{B}_\infty$ by Corollary 3.14, and each term in the first square bracket is in $\mathcal{B}(\mathcal{H})$: $R_z WJ \in \mathcal{B}(\mathcal{H})$ and $R_z\Phi P_m \in \mathcal{B}(\mathcal{H})$ by Lemma 2.38, whereas $(I + |\underline{Q}|)^\kappa(\Delta J) \in \mathcal{B}(\mathcal{H})$ and $(I + |\underline{Q}|)^\kappa J_{,m} \in \mathcal{B}(\mathcal{H})$ since the derivatives of j have compact support, so that e.g. $(\Delta j)(\underline{x}) \cdot (1 + |\underline{x}|)^\kappa \in L^\infty(\mathbb{R}^\nu)$. □

We now turn to the remaining conditions (C1)-(C3) which involve the operators $D_\pm$. Roughly speaking, $D_\pm$ will be as follows. D_+ picks out that part of a vector f in which the component of the velocity along the direction of the position vector of the particle is positive, whereas D_- picks out that part of f in which this component is negative. One can then show that, for each $\varphi \in C_{oo}^\infty(\mathbb{R})$,

$$\| (I + |\underline{Q}|)^{-\kappa} \varphi(H_o) U_t^o D_\pm \| \leq c(\varphi,\kappa)(1 + |t|)^{-\kappa} \text{ for } t \gtrless 0. \tag{5.76}$$

Thus the norm on the l.h.s. is an integrable function of t on $[0,\pm\infty)$ provided that $\kappa > 1$, which allows one to obtain (C3) for potentials decreasing more rapidly than $|\underline{x}|^{-1}$ at infinity.

Before giving a precise definition of $D_\pm$, we find it instructive to show that the analogue of the inequalities (5.76) is satisfied in classical mechanics. We consider the case $t > 0$.

In classical mechanics the state of a point particle at a fixed time is described by a pair $(\underline{x},\underline{v})$, where $\underline{x}\in\mathbb{R}^\nu$ is the position and $\underline{v}\in\mathbb{R}^\nu$ the velocity of the particle. The classical analogue of the Hilbert space norm $\|\cdot\|$ is a supremum over all $(\underline{x},\underline{v})\in\mathbb{R}^\nu\times\mathbb{R}^\nu$. The analogue of the free evolution operator U_t^0 is the correspondence $(\underline{x},\underline{v}) \mapsto (\underline{x}+\underline{v}t,\underline{v})$. The operator D_+ corresponds to the restriction $\underline{v}\cdot\underline{x} \geq 0$, and the operator $\varphi(\underline{P}^2)$, $\varphi\in C_{00}^\infty(\mathbb{R})$, imposes a minimal velocity v_{min} and a maximal velocity v_{max} (depending on φ), since the momentum is just a multiple of the velocity ($\underline{p} = m\underline{v}$, where m is the mass of the particle, which we have chosen to be $m = \frac{1}{2}$ in these notes). Thus $\varphi(\underline{P}^2)$ introduces the restriction $0 < v_{min} \leq |\underline{v}| \leq$ $\leq v_{max} < \infty$. It will be seen that the strict positivity of v_{min} is crucial.

Thus, in the evaluation of the classical analogue of the norm in (5.76), the supremum will be taken only over pairs $(\underline{x},\underline{v}_{adm})$, where $\underline{x}\in\mathbb{R}^\nu$ and $\underline{v}_{adm}$ is an <u>admissible velocity</u> relative to $\underline{x}$, i.e. satisfying $\underline{v}_{adm}\cdot\underline{x} \geq 0$ and $|\underline{v}_{adm}|\in[v_{min},v_{max}]$.

Suppose now that the particle position at $t = 0$ is $\underline{x}\in\mathbb{R}^\nu$. Then at some *positive* time t, the particle (moving with an arbitrary admissible velocity relative to $\underline{x}$) will be localized at some point in the spherical half-shell $\{\underline{x}+\underline{v}_{adm}t\}$ centered at the point $\underline{x}$ (see the figure on the next page for the case $\nu = 2$). Notice that $|\underline{x}+\underline{v}_{adm}t| \geq v_{min}t$ for all admissible velocities and each $\underline{x}\in\mathbb{R}^\nu$.

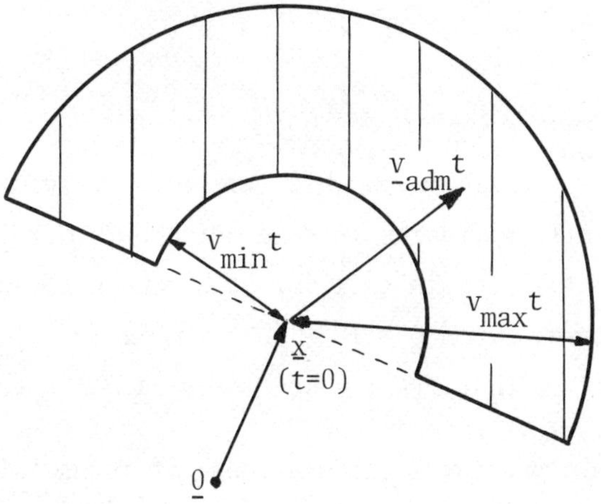

Consequently $(1+|\underline{x}+\underline{v}_{adm}t|)^{-\kappa} \leq (1+v_{min}t)^{-\kappa}$ for all $(\underline{x},\underline{v}_{adm})$. This means that, in the classical analogy (symbolized by ~) :

$$\|(I+|\underline{Q}|)^{-\kappa}\varphi(H_0)U_t^0 D_+\| \sim \sup_{(\underline{x},\underline{v}_{adm})} (1+|\underline{x}+\underline{v}_{adm}t|)^{-\kappa}|\varphi(\tfrac{1}{4}\underline{v}_{adm}^2)|$$

$$\leq (1+v_{min}t)^{-\kappa}\|\varphi\|_\infty, \qquad \forall t \geq 0,$$

which implies (5.76) for the upper sign, since $v_{min} > 0$. Similarly one may deduce the classical analogue of (5.76) for the lower sign.

We now give a precise definition of the operators $D_\pm$ in quantum mechanics. There are various ways of obtaining a quantum-mechanical analogue of the condition $\underline{x}\cdot\underline{v} > 0$. One possibility is to use the self-adjoint operator $B = \frac{1}{2}(\underline{P}\cdot\underline{Q}+\underline{Q}\cdot\underline{P})$ and to define $D_\pm$ as the projections onto the subspaces $\mathcal{H}_\pm(B)$ on which B is positive or negative ($D_\pm$ are then the so-called spectral projections of the self-adjoint operator B corresponding to the intervals $(0,\pm\infty)$). These operators were used in [12], [13].

We find it more convenient to use instead of B the following operator [14]

$$A_o = \frac{1}{2}\left(\frac{\underline{P}}{\underline{P}^2}\cdot\underline{Q} + \underline{Q}\cdot\frac{\underline{P}}{\underline{P}^2}\right), \tag{5.77}$$

with domain $D(A_o) = \tilde{C}_o^\infty(\mathbb{R}^\nu\backslash\{\underline{0}\})$, the set of all functions from $\mathbb{R}^\nu$ to $\mathbb{C}$ the Fourier transforms of which are infinitely differentiable, of bounded support and vanish in some neighbourhood of $\underline{k} = 0$. A_o is symmetric on this domain (however it has no self-adjoint extension). Furthermore, since $\underline{Q} = i\underline{\nabla}$ in $\tilde{L}^2(\mathbb{R}^\nu)$, we see that A_o leaves $D(A_o)$ invariant, so that $D(A_o) = D(A_o^n)$ for each $n \geq 1$.

In the above and later calculations, it is useful to have at our disposal the following commutation relation : if $\psi, f \in S(\mathbb{R}^\nu)$, then

$$Q_m\psi(\underline{P})f = \psi(\underline{P})Q_m f + i\frac{\partial\psi}{\partial k_m}(\underline{P})f. \tag{5.78}$$

This is obtained by writing $Q_m = i\partial/\partial k_m$ in the momentum variables and using the rule for differentiating a product.

We now record some additional relations involving A_o. For this it is useful to introduce spherical polar coordinates $(k,\underline{\omega})$ in momentum space. $k\in(0,\infty)$ is defined as $k = |\underline{k}|$, whereas $\underline{\omega} = \underline{k}/k$ is a vector on the unit sphere $S^{\nu-1}$ of $\mathbb{R}^\nu$: $S^{\nu-1} := \{\underline{k}\in\mathbb{R}^\nu \,|\, |\underline{k}| = 1\}$. Instead of k we shall also use the variable $\lambda = k^2$. (λ is called the <u>kinetic energy</u> associated with the wave vector $\underline{k}$). Using the relations $\underline{Q} = i\underline{\nabla}_k$, $\partial k/\partial k_m = k_m/k$ and $k\partial/\partial k = \sum_{m=1}^{\nu} k_m\partial/\partial k_m$, we get for $f\in D(A_o)$:

$$\begin{aligned}(FA_o f)(\underline{k}) &= \frac{i}{2}\sum_{m=1}^{\nu}\left[\frac{k_m}{k^2}\frac{\partial}{\partial k_m}\tilde{f}(\underline{k}) + \frac{\partial}{\partial k_m}\left\{\frac{k_m}{k^2}\tilde{f}(\underline{k})\right\}\right] \\ &= \frac{i}{2}\sum_{m=1}^{\nu}\left[2\frac{k_m}{k^2}\frac{\partial}{\partial k_m}\tilde{f}(\underline{k}) + \frac{k^2 - 2k_m^2}{k^4}\tilde{f}(\underline{k})\right] \\ &= i\left(\frac{1}{k}\frac{\partial}{\partial k} + \frac{\nu - 2}{2k^2}\right)\tilde{f}(k\underline{\omega}).\end{aligned} \tag{5.79}$$

In particular we have

$$A_0 = (\underline{P}^2)^{-1}[\underline{P}\cdot\underline{Q} + \frac{i(\nu-2)}{2}]. \qquad (5.80)$$

In the same way one may obtain an expression for A_0^n having all momentum operators on the left of the position operators :

$$A_0^n = (\underline{P}^2)^{-n}(\alpha_0^{(n)} + \sum_{k=1}^{n} \alpha_k^{(n)} \sum_{m_1 \ldots m_k=1}^{\nu} P_{m_1} \cdots P_{m_k} Q_{m_1} \cdots Q_{m_k}), \qquad (5.81)$$

where $\alpha_k^{(n)}$ $(k=0,\ldots,n)$ are constants depending on ν. For $n = 1$, (5.81) is identical with (5.80), and for general n it can easily be deduced from (5.80) by induction, by a calculation similar to that in (5.79).

The expression (5.79) for A_0 becomes even simpler if we map the Hilbert space $\mathcal{H} = L^2(\mathbb{R}^\nu)$ by means of a unitary transformation $\mathcal{U}_0$ onto the Hilbert space $\mathcal{K} = L^2((0,\infty), L^2(S^{\nu-1}))$. This transformation $\mathcal{U}_0$ will also be important in Chapter 6. We first give the relevant definitions. $L^2(S^{\nu-1})$ is the set of all measurable and square-integrable functions from the unit sphere $S^{\nu-1}$ to $\mathbb{C}$, with respect to Lebesgue measure on $S^{\nu-1}$. We denote this measure by $d\omega$, the scalar product in $L^2(S^{\nu-1})$ by $(\cdot,\cdot)_0$ and the norm in this space by $||\cdot||_0$.

$L^2((0,\infty), L^2(S^{\nu-1}))$ is the set of all (measurable) functions defined on $(0,\infty)$ with values in the Hilbert space $L^2(S^{\nu-1})$, whose $L^2(S^{\nu-1})$-norms are square-integrable with respect to Lebesgue measure $d\lambda$ on $(0,\infty)$. We denote the value of such a function g at the point $\lambda\in(0,\infty)$ by g_λ. Thus g_λ is a vector in $L^2(S^{\nu-1})$, i.e. a square-integrable function $g_\lambda(\underline{\omega})$ of the variable $\underline{\omega}\in S^{\nu-1}$, and $\int_0^\infty d\lambda ||g_\lambda||_0^2 < \infty$. The scalar product between $f = \{f_\lambda\}$ and $g = \{g_\lambda\}$ in $L^2((0,\infty), L^2(S^{\nu-1}))$ is

$$(f,g) = \int_0^\infty (f_\lambda, g_\lambda)_o d\lambda = \int_0^\infty d\lambda \int_{S^{\nu-1}} d\omega\, \overline{f_\lambda(\underline{\omega})} g_\lambda(\underline{\omega}). \tag{5.82}$$

Let us now define, for $f \in L^2(\mathbb{R}^\nu)$:

$$(\mathcal{U}_o f)_\lambda(\underline{\omega}) = 2^{-1/2} \lambda^{(\nu-2)/4} \tilde{f}(\lambda^{1/2}\underline{\omega}). \tag{5.83}$$

Using the fact that $d^\nu k = k^{\nu-1} dk d\omega = \frac{1}{2}\lambda^{(\nu-2)/2} d\lambda d\omega$, we see that

$$\int_0^\infty d\lambda \|(\mathcal{U}_o f)_\lambda\|_o^2 = \frac{1}{2} \int_0^\infty \lambda^{(\nu-2)/2} d\lambda \int_{S^{\nu-1}} d\omega |\tilde{f}(\lambda^{1/2}\underline{\omega})|^2$$

$$= \int_{\mathbb{R}^\nu} d^\nu k |\tilde{f}(\underline{k})|^2 = \|f\|^2.$$

Thus $\mathcal{U}_o$ is an isometric operator from $L^2(\mathbb{R}^\nu)$ to $L^2((0,\infty), L^2(S^{\nu-1}))$. In fact $\mathcal{U}_o$ is onto, hence unitary, since its inverse is easily seen to be as follows : If $g = \{g_\lambda\} \in L^2((0,\infty), L^2(S^{\nu-1}))$, then

$$[F\mathcal{U}_o^{-1} g](\underline{k}) = 2^{1/2} k^{(2-\nu)/2} g_{\underline{k}^2}(\underline{\omega}).$$

We also notice the following relation which may be deduced from Problem 4.38 and will also be proved in Lemma 6.3 :

$$[\mathcal{U}_o \varphi(H_o) f]_\lambda = \varphi(\lambda)(\mathcal{U}_o f)_\lambda. \tag{5.84}$$

This means that the unitary mapping $\mathcal{U}_o$ diagonalizes the operator $H_o = \underline{P}^2$. It is referred to as the <u>spectral transformation</u> for H_o.

Let us now calculate $\mathcal{U}_o A_o \mathcal{U}_o^{-1}$, using (5.79) and $k^{-1}\partial/\partial k = 2\partial/\partial\lambda$. If g is as above and such that $\mathcal{U}_o^{-1} g \in D(A_o)$, we obtain

$$(\mathcal{U}_o A_o \mathcal{U}_o^{-1} g)_\lambda = i\lambda^{(\nu-2)/4}[2\frac{\partial}{\partial\lambda} + \frac{\nu-2}{2\lambda}]\lambda^{-(\nu-2)/4} g_\lambda = 2i\frac{\partial}{\partial\lambda} g_\lambda ,$$

hence formally

$$\mathcal{U}_o A_o \mathcal{U}_o^{-1} = 2i\frac{\partial}{\partial\lambda}. \tag{5.85}$$

It is now convenient to view $L^2((0,\infty), L^2(S^{\nu-1}))$ as a subspace of $L^2(\mathbb{R},L^2(S^{\nu-1}))$ and to make a one-dimensional inverse Fourier transformation in this latter space with respect to the variable λ. Thus we define, for $z\in\mathbb{R}$:

$$[\mathcal{F}_1^{-1}\{f_\lambda\}]_z(\underline{\omega}) \equiv \hat{f}_z(\underline{\omega}) = (2\pi)^{-1/2} \int_{-\infty}^{\infty} e^{i\lambda z} f_\lambda(\underline{\omega})d\lambda.$$

This transforms $id/\partial\lambda$ into multiplication by z : If $g = \{\hat{g}_z\}$, then

$$(\mathcal{F}_1^{-1}\mathcal{U}_o A_o \mathcal{U}_o^{-1}\mathcal{F}_1 g)_z = 2z\hat{g}_z.$$

The operators D_+ and D_- should now select the positive and negative z-components of $\hat{g}_z$ respectively. We thus define two projections $\hat{F}_\pm$ in the z-representation of $L^2(\mathbb{R},L^2(S^{\nu-1}))$ by

$$(\hat{F}_+ g)_z = \chi_{(0,\infty)}(z)\hat{g}_z \ , \quad (\hat{F}_- g)_z = \chi_{(-\infty,0)}(z)\hat{g}_z.$$

Clearly $\hat{F}_+ + \hat{F}_- = I$. However, if $g = \{\hat{g}_z\}$ is the image under $\mathcal{F}_1^{-1}$ of a vector in $L^2((0,\infty), L^2(S^{\nu-1}))$ (i.e. having λ-support in $(0,\infty)$), then $\hat{F}_\pm g$ will not have this property. In other words, $F_\pm \equiv \mathcal{F}_1\hat{F}_\pm\mathcal{F}_1^{-1}$ do not commute with the projection E_+ of $L^2(\mathbb{R}, L^2(S^{\nu-1}))$ onto $L^2((0,\infty), L^2(S^{\nu-1}))$. To obtain operators $D_\pm$ acting in $L^2((0,\infty), L^2(S^{\nu-1}))$ (in the λ-variable), we must project back into $L^2((0,\infty), L^2(S^{\nu-1}))$. We thus take, in $L^2(\mathbb{R}^\nu)$:

$$D_\pm = \mathcal{U}_o^{-1}E_+\mathcal{F}_1\hat{F}_\pm\mathcal{F}_1^{-1}\mathcal{U}_o. \qquad (5.86)$$

Since $\hat{F}_+ + \hat{F}_- = I$, $D_+ + D_-$ is the identity operator on $L^2(\mathbb{R}^\nu)$. $D_\pm$ are self-adjoint and in $\mathcal{B}(\mathcal{H})$. In fact $\|D_\pm\| \le 1$, since $D_\pm$ are products of unitary operators and projections. Furthermore $D_\pm$ satisfy the inequality (5.76); however we only prove a slightly weaker result :

Lemma 5.33 : Let $\kappa > 0$. Then, for each $\kappa'\in[0,\kappa)$ and each $\varphi\in C_{oo}^\infty(\mathbb{R})$:

$$\|(I+|\underline{Q}|)^{-\kappa}\varphi(H_o)U_t^o D_+\| \le c(\varphi,\kappa,\kappa')t^{-\kappa'} \quad \forall t \ge 1 \qquad (5.87)$$

and

$$\|(I+|\underline{Q}|)^{-\kappa}\varphi(H_o)U_t^o D_-\| \le c(\varphi,\kappa,\kappa')|t|^{-\kappa'} \quad \forall t \le -1. \qquad (5.88)$$

<u>Proof</u> : We prove (5.87). The proof of (5.88) is similar.

(i) Choose an integer n such that $\kappa_o := n\kappa(n+\kappa)^{-1} \ge \kappa'$ and define $\alpha = \kappa_o\kappa^{-1}$. Notice that $\alpha n + \kappa_o = n$.

Let $t \ge 1$. Then $1 + t^\beta \le 2t^\beta$ for any $\beta \ge 0$. Let F_r be the projection with range $L^2(B_r)$, defined after Eq. (5.1). We may then write, for any $g \in \mathcal{H}$, and using (2.24) :

$$\|(I+|\underline{Q}|)^{-\kappa}g\| \le \|(I+|\underline{Q}|)^{-\kappa}F_{t^\alpha}g\| + \|(I+|\underline{Q}|)^{-\kappa}(I-F_{t^\alpha})g\|$$

$$\le \|F_{t^\alpha}g\| + \sup_{|\underline{x}|\ge t^\alpha}(1+|\underline{x}|)^{-\kappa}\|g\|. \qquad (5.89)$$

The square of the first term on the r.h.s. is the integral of $|g(\underline{x})|^2$ over the domain $|\underline{x}| \le t^\alpha$. In this domain we have $1 \le (1+t^\alpha)(1+|\underline{x}|)^{-1} \le 2t^\alpha(1+|\underline{x}|)^{-1}$, hence also $1 \le 2^{2n}t^{2\alpha n}\cdot(1+|\underline{x}|)^{-2n}$. Consequently

$$\|F_{t^\alpha}g\| \le 2^n t^{\alpha n}\|(I+|\underline{Q}|)^{-n}g\| = 2^n t^{-\kappa_o}t^n\|(I+|\underline{Q}|)^{-n}g\|.$$

Inserting this into (5.89), we obtain

$$\|(I+|\underline{Q}|)^{-\kappa}g\| \le 2^n t^{-\kappa_o}t^n\|(I+|\underline{Q}|)^{-n}g\| + t^{-\alpha\kappa}\|g\|$$

$$= t^{-\kappa_o}(2^n t^n\|(I+|\underline{Q}|)^{-n}g\| + \|g\|).$$

Taking $g = U_t^o\varphi(H_o)D_+f$, we see from the preceding inequality that, for $t \ge 1$:

$$\|(I+|\underline{Q}|)^{-\kappa}U_t^o\varphi(H_o)D_+\|$$

$$\leq t^{-\kappa_0}(2^n t^n\|(1+|\underline{Q}|)^{-n}U_t^o\varphi(H_o)D_+\| + \|U_t^o\varphi(H_o)D_+\|).$$

Since $\kappa_o \geq \kappa'$ and $\|U_t^o\varphi(H_o)D_+\| \leq (2\pi)^{-1/2}\|\tilde{\varphi}\|_1$, (5.87) will be proven if we can show that, for each $n \geq 1$ and each $\varphi \in C_{oo}^\infty(\mathbb{R})$, there is a number $M_n(\varphi) < \infty$ such that, for all $t \geq 1$:

$$t^n\|D_+\varphi(H_o)U_t^{o*}(I+|\underline{Q}|)^{-n}\| \equiv t^n\|(I+|\underline{Q}|)^{-n}U_t^o\bar{\varphi}(H_o)D_+\| \leq M_n(\varphi). \quad (5.90)$$

(ii) To prove (5.90), we choose $\psi \in C_{oo}^\infty(\mathbb{R})$ such that $\psi(\lambda)\varphi(\lambda) = \varphi(\lambda)$ (i.e. $\psi(\lambda) = 1$ for λ in the support of φ). We set $W_t =$ $= D_+\varphi(H_o)U_t^{o*}(I+|\underline{Q}|)^{-n}$. From the definition (5.86) of D_+ we then get that

$$t^n\|W_t f\| \leq t^n\|\hat{F}_+ F_1^{-1} U_o \varphi(H_o)U_t^{o*}(I+|\underline{Q}|)^{-n}f\|,$$

where the last norm is in $L^2(\mathbb{R}, L^2(S^{\nu-1}))$ with respect to the variable z. We assume $f \in S(\mathbb{R}^\nu)$. Then $(I+|\underline{Q}|)^{-n}f \in S(\mathbb{R}^\nu)$, hence $h :=$ $:= \varphi(H_o)(I+|\underline{Q}|)^{-n}f \in D(A_o) = D(A_o^m)$ for each $m = 1,2,\ldots$. We shall show in (iii) and (iv) below that

$$t^n\|\hat{F}_+ F_1^{-1} U_o U_t^{o*} h\| \leq 2^{-n}\|\hat{F}_+ F_1^{-1} U_o U_t^{o*}\psi(H_o)A_o^n h\| \quad (5.91)$$

and $\psi(H_o)A_o^n\varphi(H_o)(I+|\underline{Q}|)^{-n} \in B(H)$. (5.92)

It follows that

$$t^n\|W_t f\| \leq 2^{-n}\|\psi(H_o)A_o^n\varphi(H_o)(I+|\underline{Q}|)^{-n}\|\,\|f\|,$$

for each f in the dense set $S(\mathbb{R}^\nu)$, which implies (5.90).

(iii) To prove (5.91), we notice that $h = \psi(H_o)h$. We write the norms in (5.91) as integrals over the variable z, use the fact that $t^{2n} \leq (t+z)^{2n}$ if $t \geq 1$ and $z \geq 0$ and then integrate by parts to obtain that

$$t^{2n}\|\hat{F}_+F_1^{-1}\mathcal{U}_o U_t^{o*}h\|^2$$

$$= (2\pi)^{-1}t^{2n}\int_0^\infty dz\|\int_{-\infty}^\infty e^{i(z+t)\lambda}\psi(\lambda)(\mathcal{U}_oh)_\lambda d\lambda\|_o^2$$

$$\leq (2\pi)^{-1}\int_0^\infty dz\|(t+z)^n\int_{-\infty}^\infty e^{i(z+t)\lambda}\psi(\lambda)(\mathcal{U}_oh)_\lambda d\lambda\|_o^2$$

$$= (2\pi)^{-1}\int_0^\infty dz\|\int_{-\infty}^\infty [\frac{d^n}{d\lambda^n}e^{i(z+t)\lambda}]\psi(\lambda)(\mathcal{U}_oh)_\lambda d\lambda\|_o^2$$

$$= (2\pi)^{-1}\int_0^\infty dz\|\int_{-\infty}^\infty e^{i(z+t)\lambda}\frac{d^n}{d\lambda^n}[\psi(\lambda)(\mathcal{U}_oh)_\lambda]d\lambda\|_o^2. \qquad (5.93)$$

Using the fact that $\psi(\lambda) = 1$ on the support of $(\mathcal{U}_oh)_\lambda$ and then (5.85), we find

$$\frac{d^n}{d\lambda^n}[\psi(\lambda)(\mathcal{U}_oh)_\lambda] = \psi(\lambda)\frac{d^n}{d\lambda^n}(\mathcal{U}_oh)_\lambda = (2i)^{-n}\psi(\lambda)(\mathcal{U}_oA_o^nh)_\lambda.$$

This shows that the last expression in (5.93) is equal to $2^{-2n}\|\hat{F}_+F_1^{-1}\mathcal{U}_oU_t^{o*}\psi(H_o)A_o^nh\|^2$, which proves (5.91).

(iv) It remains to verify (5.92). From (5.81) we get

$$\psi(H_o)A_o^n\varphi(H_o)(I+|\underline{Q}|)^{-n} = \alpha_o^{(n)}H_o^{-n}\varphi(H_o)(I+|\underline{Q}|)^{-n} +$$

$$+\sum_{k=1}^n \alpha_k^{(n)}\sum_{m_1\ldots m_k=1}^{\nu}[H_o^{-n}P_{m_1}\cdots P_{m_k}\psi(H_o)]Q_{m_1}\cdots Q_{m_k}\varphi(H_o)(I+|\underline{Q}|)^{-n}.$$

$H_o^{-n}\varphi(H_o)$ and $H_o^{-n}P_{m_1}\cdots P_{m_k}\psi(H_o)$ are multiplication operators in $\tilde{L}^2(\mathbb{R}^\nu)$ by functions belonging to $C_o^\infty(\mathbb{R}^\nu)$, see Problem 4.38, hence these operators belong to $\mathcal{B}(L^2(\mathbb{R}^\nu))$ by Proposition 2.12. (Notice that it is at this point that we use the hypothesis $\varphi(\lambda) = 0$ near $\lambda = 0$ and $\lambda = \infty$.) Also, from the commutation relation (5.78), one sees that $Q_{m_1}\cdots Q_{m_k}\varphi(H_o)(I+|\underline{Q}|)^{-n}$ is a finite sum of terms of the

$$\chi_{s_1\ldots s_i}(\underline{P})Q_{s_1}\cdots Q_{s_i}(I+|\underline{Q}|)^{-n}.$$

Here $\chi_{s_1 \ldots s_i}$ is in $C_0^\infty(\mathbb{R}^\nu)$, since it is obtained from φ by a finite number of partial derivations, and $i \le k$. Since $k \le n$, we have $Q_{s_1} \cdots Q_{s_i}(I + |\underline{Q}|)^{-n} \in \mathcal{B}(\mathcal{H})$, and clearly $\chi_{s_1 \ldots s_i}(\underline{P}) \in \mathcal{B}(\mathcal{H})$. This proves (5.92). □

We shall now discuss (C2) and (C3), using Lemma 5.33. (C2) involves only the free Hamiltonian and can be obtained as follows : Let $f \in D(I + |\underline{Q}|)$, i.e. $f = (I + |\underline{Q}|)^{-1} g$ for some $g \in \mathcal{H}$, and let $\varphi \in C_{oo}^\infty(\mathbb{R})$. Then, since $U_t^{o*} = U_{-t}^o$:

$$\|D_- U_t^o \varphi(H_o) f\| \le \|D_- U_t^o \varphi(H_o)(I + |\underline{Q}|)^{-1}\| \, \|g\|$$

$$= \|(I + |\underline{Q}|)^{-1} \overline{\varphi}(H_o) U_{-t}^o D_-\| \, \|g\|, \tag{5.94}$$

which converges to zero as $t \to +\infty$ (i.e. $-t \to -\infty$) by (5.88). Since $D(I + |\underline{Q}|)$ is dense in $\mathcal{H}$, (5.94) shows that $\text{s-lim}\, D_- U_t^o h = 0$ as $t \to +\infty$ for all h in the total subset $\{\varphi(H_o) D(I + |\underline{Q}|) \,|\, \varphi \in C_{oo}^\infty(\mathbb{R})\}$, see Proposition 4.27. Therefore, by Lemma 1.8, $\text{s-lim}\, D_- U_t^o = 0$ as $t \to +\infty$. This proves (C2).

As regards (C3), we must assume that $v \in V_{\Gamma,\kappa}$ with $\kappa > 1$. We know from (5.73) and (5.75) that $Y_z = B_z (I + |\underline{Q}|)^{-\kappa} R_z^o$, where B_z is some operator in $\mathcal{B}(\mathcal{H})$. If $\varphi \in C_{oo}^\infty(\mathbb{R})$, we define ψ by $\psi(\lambda) = (\lambda - z)^{-1} \varphi(\lambda)$ and notice that $\psi \in C_{oo}^\infty(\mathbb{R})$ also. Consequently, by (5.87),

$$\|Y_z \varphi(H_o) U_t^o D_+\| \le \|B_z\| \, \|(I + |\underline{Q}|)^{-\kappa} \psi(H_o) U_t^o D_+\|$$

$$\le c(\psi, \kappa, \kappa') \|B_z\| (1 + t)^{-\kappa'} \qquad \forall t \ge 0.$$

Since $\kappa > 1$, we may take $\kappa' > 1$, which shows that $\|Y_z \varphi(H_o) U_t^o D_+\|$ is integrable on $[0, +\infty)$. This proves (C3).

We shall now summarize the results obtained so far, using

Propositions 5.28 and 5.29.

Proposition 5.34 : Assume that $v \in V_{\emptyset,\kappa}$ for some $\kappa > 1$, i.e. $v(\underline{x}) = [w_1(\underline{x}) + w_2(\underline{x})](1+|\underline{x}|)^{-\kappa}$ with $w_1 \in L^{\infty}(\mathbb{R}^{\nu})$ and $w_2 \in L^p(\mathbb{R}^{\nu})$, $p \geq 2$, $p > \nu/2$. Let $H = H_o + V$. Then the wave operators $\Omega_{\pm} = \text{s-lim}\, U_t^* U_t^o$ as $t \to \pm\infty$ exist and are complete in the sense that

$$\mathcal{R}(\Omega_{\pm}) = \mathcal{H}_p(H)^{\perp} = \mathcal{H}_{ac}(H) = \mathcal{M}_{\infty}^{\pm}(H). \tag{5.95}$$

The operator H has no singularly continuous spectrum, each non-zero eigenvalue of H is of finite multiplicity, and the only possible accumulation points of these eigenvalues are $\lambda = 0$ and $\lambda = \pm\infty$ (see also Remark 5.36 (a)).

Proposition 5.35 : (a) Let Γ be a bounded closed set of measure zero in $\mathbb{R}^{\nu}$. Assume that $v \in V_{\Gamma,\kappa}$ for some $\kappa > 1$. Let H be an arbitrary self-adjoint extension of the minimal operator $\hat{H}$ given in (2.54), and U_t the associated evolution group. Then the wave operators $\Omega_{\pm} = \text{s-lim}\, U_t^* U_t^o$ as $t \to \pm\infty$ exist.

(b) Assume in addition that, for each $f \in \mathcal{H}_c(H)$,

$$\lim_{T\to\infty} \frac{1}{T} \int_0^{\pm T} \|F_r U_t f\|^2 dt = 0, \tag{5.96}$$

where r is such that Γ is contained in the interior of the ball $B_r = \{\underline{x} \in \mathbb{R}^{\nu} \,|\, |\underline{x}| \leq r\}$. Then $\Omega_{\pm}$ are complete, i.e. one has (5.95). In particular H has no singularly continuous spectrum.

Proof : It suffices to remark that (5.96) implies (C7) with $K_{\pm} = I$ (if $r \geq M + 2$, one may take $J_+ = J_- = j(\underline{Q})$ with j as indicated earlier; if $r < M + 2$, a slightly modified j has to be taken). The fact that $\mathcal{H}_p(H)^{\perp} = \mathcal{M}_{\infty}^{\pm}(H)$ follows from the completeness of $\Omega_{\pm}$, Propositions 5.2 and 5.14 (f) and Example 5.6. □

Remark 5.36 : (a) Under the assumptions of Proposition 5.34 it is possible to show by different methods that the eigenvalues of H form a bounded set, hence that they can accumulate at most at $\lambda = 0$; for the negative eigenvalues, this result follows from Propositions 2.23 and 2.28.

(b) Proposition 5.35 (a) shows that local singularities of the potential have no influence on the *existence* of the wave operators $\Omega_\pm$. Its behaviour at infinity however is crucial. It is seen that our existence proof breaks down when $\kappa \leq 1$, since we then cannot verify (C3). In fact it is known that $\{U_t^* U_t^o\}$ is not strongly convergent for Coulomb and other long range potentials [15]. However one then has wave operators $W_\pm = \text{s-lim}\ U_t^* J_\pm U_t^o$ as $t \to \pm\infty$ for suitable operators $J_\pm$ (depending on the potential) [16]-[18]. The completeness of such wave operators has also been shown [18], though by methods that are different from those used here.

(c) As far as the *completeness* of $\Omega_\pm$ is concerned, it is seen from Proposition 5.35 (b) (and also from the discussion at the end of Section 5.1) that it depends on the local singularities of v. If v is only weakly singular, $\Omega_\pm$ are always complete. If v has strong local singularities, then a necessary and sufficient condition for completeness is (5.96), stating that all vectors in $\mathcal{H}_c(H)$ are evanescent on the time average from some ball containing all points of strong singularity of v in its interior. If one drops the hypothesis (5.96), one can prove some forms of generalized asymptotic completeness. Namely one has generalized asymptotic completeness on the average as well as asymptotic completeness in the geometric sense at positive and at negative times :

Proposition 5.37 : Under the hypotheses of Proposition 5.35 (a), one has (5.49) and (5.47), i.e.

$$\mathcal{H}_c(H) = \mathcal{R}(\Omega_+) \oplus \overline{\mathcal{M}}_\Gamma^+(H) = \mathcal{R}(\Omega_-) \oplus \overline{\mathcal{M}}_\Gamma^-(H) \tag{5.97}$$

and

$$\overline{M_\Gamma^\pm}(H) \subseteq M_o^\pm(H). \tag{5.98}$$

One also has

$$\mathcal{H} = M_\infty^+(H) \oplus M_o^+(H) = M_\infty^-(H) \oplus M_o^-(H). \tag{5.99}$$

We add that, by using the spectral theorem, one can show furthermore that [10]

$$\mathcal{H}_w(H) = \mathcal{R}(\Omega_+) \oplus M_\Gamma^+(H) = \mathcal{R}(\Omega_-) \oplus M_\Gamma^-(H)$$

and

$$\mathcal{H}_{sc}(H) \subseteq M_o(H). \tag{5.100}$$

(5.100) follows from Proposition 5.37 since $\mathcal{H}_{sc}(H) \subseteq \mathcal{R}(\Omega_+)^\perp \cap \mathcal{R}(\Omega_-)^\perp$.

Proof of Proposition 5.37 : Let $\varphi \in C_\Gamma$. Then $\varphi(1-j) \in C_\Gamma \cap C_o^\infty(\mathbb{R}^\nu)$. Choose $\psi \in C_\Gamma \cap C_o^\infty(\mathbb{R}^\nu)$ such that $\psi(\underline{x}) = 1$ on the support of $\varphi(1-j)$, and denote by Ψ the multiplication operator by $\psi(\underline{x})$. Then

$$\Phi(I-J)R_z = [\Phi(I-J)R_z^o][(H_o - z)\Psi R_z].$$

The first factor is in $\mathcal{B}_\infty$ by Corollary 3.14, the second one in $\mathcal{B}(\mathcal{H})$ by Proposition 2.34. Hence, using Proposition 4.25, we see that (C7) holds with $K_\pm = \Phi = \varphi(\underline{Q})$. Thus, by Lemma 5.27,

$$\lim_{T\to\infty} \frac{1}{T} \int_0^{\pm T} \|\varphi(\underline{Q})U_t f\|^2 dt = 0$$

for each $\varphi \in C_\Gamma$ and each $f \in \mathcal{H}_c(H) \cap \mathcal{R}(\Omega_\pm)^\perp$. This shows that $\mathcal{H}_c(H) \cap \mathcal{R}(\Omega_\pm)^\perp \subseteq \overline{M_\Gamma^\pm}$. (5.97) now follows by virtue of Propositions 5.14 (f) and 5.10 (c).

For the proof of (5.98), we refer to [10, Theorem 2]. (5.99) is an immediate consequence of (5.97) and (5.98), since $\mathcal{R}(\Omega_\pm) \subseteq M_\infty^\pm(H)$ and $\mathcal{H}_p(H) \subseteq M_o(H)$. □

CHAPTER 6 : SCATTERING THEORY

Scattering theory is the mathematical description of the following type of physical experiment : A beam of particles with well-defined initial conditions is made to collide with a fixed target, and one observes the angular distribution as well as other physical properties of the scattered particles after the collision.

By using the wave operators, one can predict the results of such experiments if one assumes the interaction between the constituents of the beam and those of the target to be known. For this, one divides the problem into two parts by noticing that the beam as well as the target are made up of a large number of identical particles or atoms :

i) One describes the scattering of a single particle of the beam by a single scattering center (i.e. a single constituent or atom of the target). In the simplest case this is a two-body problem; in more complex situations, when the constituents of the beam or the target are composite systems such as atoms, one has an N-body problem, where N is however *small* (for example, in proton-proton scattering one has N = 2, in proton-deuteron scattering one has N = 3, since the deuteron is composed of a proton and a neutron).

ii) By taking into account the properties of the beam and of the target, one treats the real situation - involving a very large number of particles - by using the results of i) and an appropriate statistical method.

The field of scattering theory is vast, and we can only treat some basic aspects here. We shall consider the situation where N = 2 in i); if the interaction between the two particles is invariant under simultaneous translations of both particles, it is easy to reduce the problem to that of scattering of a single particle by a potential (see e.g. [AJS, Chapter 7-5]). This allows us to use the results on Schrödinger operators obtained in the preceding chapters.

Problem i) is treated in Section 6.1 where we introduce the scattering operator and deduce some of its properties. In Section 6.2 we consider part ii) of the problem and deduce an expression for the probability of scattering of a beam into a fixed cone in configuration space. This leads to the notion of the scattering cross section, some properties of which we discuss in the final Section 6.3.

6.1 The Scattering Operator and the S-matrix

Let $\{U_t\}$ and $\{U_t^o\}$ be two evolution groups, $E_\pm$ two projections commuting with each U_t^o and $J_\pm \in \mathcal{B}(\mathcal{H})$. Assuming that the wave operators $W_\pm = \text{s-lim}\, U_t^* J_\pm U_t^o E_\pm$ as $t \to \pm\infty$ exist, one may define a bounded linear operator S, called the <u>scattering operator</u>, as follows :

$$S := W_+^* W_- , \qquad D(S) = \mathcal{H}. \tag{6.1}$$

We shall first establish some mathematical properties of this operator and then comment on its interpretation in scattering theory.

<u>Proposition 6.1</u> :

(a) $SU_t^o = U_t^o S \quad \forall\, t \in \mathbb{R}.$ (6.2)

(b) If $f \in D(H_o)$, then $Sf \in D(H_o)$ and $H_o Sf = SH_o f$.

<u>Proof</u> : (6.2) follows from (5.51) :

$$SU_t^o = W_+^* W_- U_t^o = W_+^* U_t W_- = U_t^o W_+^* W_- = U_t^o S.$$

The proof of (b) is similar to that of Proposition 5.13 (c). □

Proposition 6.1 expresses the fact that the operator S does not intermix different spectral values of the free Hamiltonian H_o. If H_o had pure point spectrum, then this would mean - as in Lemma 4.7 - that S maps each eigensubspace of H_o into itself. Now scattering is associated with a continuous spectrum, and in this case there is a generalization of the concept of "leaving each eigensubspace invariant", called decomposability. We shall explain this for the case $H_o = \underline{P}^2$ in $L^2(\mathbb{R}^\nu)$, since this operator can be diagonalized explicitly. Similar results can be proved on an abstract level for a general self-adjoint operator H_o (see e.g. [AJS, Chapter 5-7]).

The spectral transformation $\mathcal{U}_o$ diagonalizing $H_o = \underline{P}^2$ was defined in Section 5.4. It is a unitary operator from $L^2(\mathbb{R}^\nu)$ onto $\mathcal{G} := L^2((0,\infty),\mathcal{H}_o)$ given by (5.83), where $\mathcal{H}_o := L^2(S^{\nu-1})$ is the Hilbert space of all square-integrable functions defined on the unit sphere $S^{\nu-1}$ of $\mathbb{R}^\nu$. The linear operator A acting in $\mathcal{G}$ is said to be <u>decomposable</u> if it has the following form : for each $\lambda > 0$ there is an operator $A(\lambda) \in \mathcal{B}(\mathcal{H}_o)$ such that

$$D(A) = \{f | \int ||A(\lambda)f_\lambda||_o^2 d\lambda < \infty\} \tag{6.3}$$

and

$$(Af)_\lambda = A(\lambda)f_\lambda . \tag{6.4}$$

Thus, after diagonalization of H_o, a decomposable operator A may be decomposed into a family $\{A(\lambda)\}$ of operators each of which acts in $\mathcal{H}_o$, in such a way that the different spectral values of H_o do not get mixed together. (The family of operators $\{A(\lambda)\}$ must also have the property of being measurable, i.e. $\forall f,g \in \mathcal{H}_o$ the function $\lambda \mapsto (f,A(\lambda)g)$ must be Lebesgue measurable.)

If A is decomposable in $\mathcal{G}$, we shall write $A = \{A(\lambda)\}$, and we denote by $\|A(\lambda)\|_o$ the norm of the operator $A(\lambda)$. Without proof we state the following simple facts about decomposable operators (the proof of (a) is similar to that of Proposition 2.12).

<u>Proposition 6.2</u> : (a) Let A be decomposable. Then A is bounded with $D(A) = \mathcal{G}$ if and only if $\|A(\lambda)\|_o$ is an essentially bounded function of λ, and

$$\|A\| = \operatorname*{ess\,sup}_{\lambda>0} \|A(\lambda)\|_o . \tag{6.5}$$

(b) If $A = \{A(\lambda)\}$ and $B = \{B(\lambda)\}$ are decomposable and in $\mathcal{B}(\mathcal{G})$, then $A + \alpha B$, AB and A^* are also decomposable and given by

$$A + \alpha B = \{A(\lambda) + \alpha B(\lambda)\}, \tag{6.6}$$

$$AB = \{A(\lambda)B(\lambda)\}, \tag{6.7}$$

$$A^* = \{A(\lambda)^*\}. \tag{6.8}$$

A special class of decomposable operators are the <u>diagonalizable</u> operators which have the form $A = \{\varphi(\lambda)I_o\}$, where φ is a function from $(0,\infty)$ to $\mathbb{C}$ and I_o denotes the identity operator in $\mathcal{H}_0$. In this case A is the multiplication operator by $\varphi(\lambda)$ in $L^2((0,\infty),\mathcal{H}_o)$. It is natural to call this operator $\varphi(H_o)$, since H_o is just multiplication by λ. Of course this definition of $\varphi(H_o)$ is equivalent to that given in Section 4.2 if φ belongs to the class of functions considered there. Indeed we have :

<u>Lemma 6.3</u> : (a) The operator $\mathcal{U}_o U_t^o \mathcal{U}_o^{-1}$ is diagonalizable and given by $\mathcal{U}_o U_t^o \mathcal{U}_o^{-1} = \{\exp(-i\lambda t)I_o\}$.

(b) Assume φ is the inverse Fourier transform of a function $\tilde{\varphi} \in L^1(\mathbb{R})$, and let $\varphi(H_o)$ be defined by (4.37). Then $\mathcal{U}_o\varphi(H_o)\mathcal{U}_o^{-1}$ is diagonalizable and given by $\mathcal{U}_o\varphi(H_o)\mathcal{U}_o^{-1} = \{\varphi(\lambda)I_o\}$.

Proof : (a) is a direct consequence of (4.93) and (5.83). To prove (b), it suffices to show, by virtue of Lemma 1.5, that

$$(f, \mathcal{U}_o \varphi(H_o) \mathcal{U}_o^{-1} g) = \int (f_\lambda, \varphi(\lambda) g_\lambda)_o d\lambda \qquad \forall\, f,g \in G. \tag{6.9}$$

Now, by (a) and (5.82),

$$(f, \mathcal{U}_o \varphi(H_o) \mathcal{U}_o^{-1} g) = (2\pi)^{-1/2} \int_{-\infty}^{\infty} \tilde{\varphi}(t)(f, \mathcal{U}_o U_{-t}^o \mathcal{U}_o^{-1} g) dt$$

$$= (2\pi)^{-1/2} \int_{-\infty}^{\infty} dt\, \tilde{\varphi}(t) \int e^{i\lambda t} (f_\lambda, g_\lambda)_o d\lambda = \int \varphi(\lambda)(f_\lambda, g_\lambda)_o d\lambda,$$

because the order of integration may be interchanged by Fubini's theorem (since $\tilde{\varphi} \in L^1(\mathbb{R})$ and $\lambda \mapsto (f_\lambda, g_\lambda)_o \in L^1(0,\infty)$). This proves (6.9). □

From Lemma 6.3 (a) we see that each decomposable operator A commutes with U_t^o :

$$(\mathcal{U}_o A U_t^o f)_\lambda = A(\lambda)[e^{-i\lambda t}(\mathcal{U}_o f)_\lambda] = e^{-i\lambda t}[A(\lambda)(\mathcal{U}_o f)_\lambda] = (\mathcal{U}_o U_t^o A f)_\lambda .$$

In the next proposition we prove the converse of this :

Proposition 6.4 : Assume that $B \in \mathcal{B}(L^2(\mathbb{R}^\nu))$ is such that $BU_t^o = U_t^o B$ for each $t \in \mathbb{R}$. Then $\mathcal{U}_o B \mathcal{U}_o^{-1}$ is a decomposable operator in G.

Proof : We shall abstain from writing the unitary operator $\mathcal{U}_o$, i.e. we think of B as acting directly in G.

(i) Let $\{e_m^o\}$ be an orthonormal basis of H_o, and set $e_{m,\lambda} = \exp(-\lambda) e_m^o$. Clearly $e_m = \{e_{m,\lambda}\}$ defines a vector in G, and for each fixed λ, the set $\{e_{m,\lambda}\}$ is total in H_o. We also denote by $\mathcal{D}_0 \subseteq H_o$ the set of all finite linear combinations of the vectors $\{e_m^o\}$ with *rational* coefficients, and by $\mathcal{D} \subseteq G$ the set of all finite linear combinations with rational coefficients of the vectors $\{e_m\}$. Notice that $f \in \mathcal{D}$ if and only if $f_\lambda = \exp(-\lambda) f_o$ for some $f_o \in \mathcal{D}_0$.

For each $\lambda > 0$, define an operator $C(\lambda)$ in $\mathcal{H}_o$ with domain $\mathcal{D}_o$ by setting

$$C(\lambda)e_m^o = \exp(\lambda)[B^*e_m]_\lambda \tag{6.10}$$

and extending this definition by linearity to $\mathcal{D}_o$. In (6.10) we choose for each m a fixed representative in the equivalence class of functions $(B^*e_m)_\lambda$ (cf. Section 1.1). A different choice would change the definition of $C(\lambda)$ at most on a countable union of sets Γ_m of measure zero, in other words on a set of measure zero. Thus (6.10) defines the operators $C(\lambda)$ for almost all λ.

(ii) Let $f\in\mathcal{D}$ and $\varphi\in C_o^\infty(\mathbb{R})$. We have $f_\lambda \in D(C(\lambda))$ for almost all λ and $\varphi(H_o)B^* = B^*\varphi(H_o)$. Thus, since $C(\lambda)e_{m,\lambda} = (B^*e_m)_\lambda$:

$$\int|\varphi(\lambda)|^2\|C(\lambda)f_\lambda\|_o^2 d\lambda = \int\|[\varphi(H_o)B^*f]_\lambda\|_o^2 d\lambda$$

$$= \|\varphi(H_o)B^*f\|^2 = \|B^*\varphi(H_o)f\|^2 \leq \|B^*\|^2\|\varphi(H_o)f\|^2$$

$$= \|B\|^2\int|\varphi(\lambda)|^2\|f_\lambda\|_o^2 d\lambda,$$

whence

$$\int|\varphi(\lambda)|^2(\|B\|^2\|f_\lambda\|_o^2 - \|C(\lambda)f_\lambda\|_o^2) \geq 0. \tag{6.11}$$

Given any finite interval $[a,b]$, we choose a sequence $\{\varphi_n\}\in C_o^\infty(\mathbb{R})$ such that $|\varphi_n(\lambda)| \leq 1$ and $\varphi_n(\lambda) \to \chi_{[a,b]}(\lambda)$ as $n \to \infty$ for each λ. One then obtains from (6.11) and the Lebesgue dominated convergence theorem that

$$\int_a^b(\|B\|^2\|f_\lambda\|_o^2 - \|C(\lambda)f_\lambda\|_o^2)d\lambda$$

$$= \lim_{n\to\infty}\int|\varphi_n(\lambda)|^2(\|B\|^2\|f_\lambda\|_o^2 - \|C(\lambda)f_\lambda\|_o^2)d\lambda \geq 0. \tag{6.12}$$

Since $d/dx \int_0^x\psi(s)ds = \psi(x)$ a.e. [R, Section 5.3], (6.12) implies that

$$||C(\lambda)f_\lambda||_o \le ||B||\,||f_\lambda||_o \qquad \forall\, \lambda\notin\Delta(f), \tag{6.13}$$

where $\Delta(f)$ is a set of measure zero depending on f.

Since $\mathcal{D}$ is a countable set, the union of all $\Delta(f)$, as f varies over $\mathcal{D}$, is a set of measure zero which we denote by Δ. Thus (6.13) implies that $C(\lambda)$ is bounded for all $\lambda\notin\Delta$. Hence, by Proposition 1.7, its closure $\overline{C(\lambda)}$ is in $\mathcal{B}(\mathcal{H}_o)$ and $||\overline{C(\lambda)}||_o \le ||B||$ $(\lambda\notin\Delta)$.

(iii) Now set $B(\lambda) = \overline{C(\lambda)}^* = C(\lambda)^*$ for $\lambda\notin\Delta$ and $B(\lambda) = 0$ for $\lambda\in\Delta$, and let $B' = \{B(\lambda)\}$ be the corresponding decomposable operator. We have $||B'|| = \text{ess sup}||C(\lambda)||_o \le ||B||$. To prove the proposition, it suffices to show that $B' = B$.

For this, let $g\in G$. Then

$$\int e^{-i\lambda t}(e_{m,\lambda},B(\lambda)g_\lambda)_o d\lambda = \int (C(\lambda)e_{m,\lambda},(U_t^o g)_\lambda)_o d\lambda$$

$$= \int ((B^*e_m)_\lambda,(U_t^o g)_\lambda)_o d\lambda = (B^*e_m,U_t^o g)$$

$$= (U_t^{o*}e_m,Bg) = \int e^{-i\lambda t}(e_{m,\lambda},(Bg)_\lambda)_o d\lambda.$$

This means that the Fourier transform of the L^1-function $\lambda \mapsto (e_{m,\lambda},B(\lambda)g_\lambda)_o - (e_{m,\lambda},(Bg)_\lambda)_o$ is zero. Hence this function is zero a.e. [SW, Corollary I.1.21]. In other words,

$$(e_{m,\lambda},B(\lambda)g_\lambda)_o = (e_{m,\lambda},(Bg)_\lambda)_o \quad \text{a.e.} \tag{6.14}$$

Since $\{e_{m,\lambda}\}$ is total in $\mathcal{H}_o$ for each fixed λ, we deduce from (6.14) and Lemma 1.5 that $B(\lambda)g_\lambda = (Bg)_\lambda$ a.e., which means that $B'g = Bg$ for each $g\in G$. □

It follows from Propositions 6.1 and 6.4 that the scattering operator S is decomposable : $\mathcal{U}_o S\mathcal{U}_o^{-1} = \{S(\lambda)\}$. The operator $S(\lambda)$ acting in $\mathcal{H}_o = L^2(S^{\nu-1})$ is called the S-matrix at energy λ. We shall see later that it is related in a simple way to the scattering cross section.

It is also convenient to define an operator R by

$$R := S - I. \tag{6.15}$$

Clearly R is decomposable, $\mathcal{U}_o R \mathcal{U}_o^{-1} = \{R(\lambda)\}$, and the R-matrix $R(\lambda)$ is given by $R(\lambda) = S(\lambda) - I_o$.

In many physical situations S is a unitary operator. The next proposition contains conditions for S to be unitary in the special case where $J_\pm = I$ and $E_\pm = I$. Similar results can be obtained in the general case.

Proposition 6.5 : Assume that $\Omega_\pm = \text{s-lim } U_t^* U_t^o$ as $t \to \pm\infty$ exist, and let $S = \Omega_+^* \Omega_-$. Then

(a) S is isometric if and only if $\mathcal{R}(\Omega_-) \subseteq \mathcal{R}(\Omega_+)$.

(b) S is unitary if and only if $\mathcal{R}(\Omega_-) = \mathcal{R}(\Omega_+)$. In particular S is unitary if one has asymptotic completeness in the ordinary or in the geometric sense.

Proof : We use the projections $F_\pm = \Omega_\pm \Omega_\pm^*$ defined in (5.37) and the fact that $\mathcal{R}(F_\pm) = \mathcal{R}(\Omega_\pm)$.

(a) $S^*S = (\Omega_+^*\Omega_-)^*(\Omega_+^*\Omega_-) = \Omega_-^*\Omega_+\Omega_+^*\Omega_- = \Omega_-^*F_+\Omega_-$.

If $\mathcal{R}(\Omega_-) \subseteq \mathcal{R}(\Omega_+)$, then $F_+\Omega_- = \Omega_-$, hence $S^*S = \Omega_-^*\Omega_- = I$. In this case S is isometric.

If $\mathcal{R}(\Omega_-) \not\subseteq \mathcal{R}(\Omega_+)$, there is a vector f such that $\Omega_- f \notin \mathcal{R}(\Omega_+)$, hence $\|F_+\Omega_- f\| < \|\Omega_- f\| = \|f\|$. Then $\|\Omega_-^*F_+\Omega_- f\| \le \|F_+\Omega_- f\| < \|f\|$, and consequently $S^*S \neq I$.

(b) S is unitary if and only if $S^*S = I$ and $SS^* = I$. As $S^* = \Omega_-^*\Omega_+$, one has as in (a) that $SS^* = I \Leftrightarrow \mathcal{R}(\Omega_+) \subseteq \mathcal{R}(\Omega_-)$. The result of (b) is now immediate. □

Finally we turn to the interpretation of the scattering operator. We assume that $J_\pm = I$ and $E_\pm = E_\infty^\pm(H_0)$ (cf. the beginning of Section 5.2). Let f be a vector in the subspace $M_\infty^-(H_0)$ and $g = \Omega_- f$. Then, by (5.26),

$$||U_t g - U_t^0 f|| \to 0 \quad \text{as} \quad t \to -\infty. \tag{6.16}$$

Furthermore $F_+ g \in R(\Omega_+)$, so that

$$||U_t F_+ g - U_t^0 \Omega_+^* g|| = ||U_t F_+ g - U_t^0 S f|| \to 0 \quad \text{as} \quad t \to +\infty. \tag{6.17}$$

The meaning of the preceding two relations may be described as follows. The vector $f \in M_\infty^-(H_0)$ is interpreted as the initial state of a scattering event, given *at time* $t = 0$. g is a state evolving under the real evolution group $\{U_t\}$ which is indistinguishable from the initial state f, evolving under the *free* evolution group $\{U_t^0\}$, at very large *negative* times (before the interaction becomes effective).

Now write $g = F_+ g + (I - F_+)g$. The part $F_+ g$ of g is a scattering state at $t = +\infty$, i.e. $F_+ g \in M_\infty^+(H)$ (see Corollary 5.15). Moreover, its time evolution $U_t F_+ g$ becomes indistinguishable at very large *positive* times from the state obtained by letting $Sf \in M_\infty^+(H_0)$ evolve under the free evolution group $\{U_t^0\}$. Thus Sf may be interpreted as the outgoing part of the final state (given also at time $t = 0$); this is meant in the sense of (6.17) : the free evolution of Sf at large positive times gives the outgoing part of a scattering event initiated, in the sense of (6.16), in the state $f \in M_\infty^-(H_0)$.

The advantage of this representation is that one may describe the outgoing state at large positive times by using the scattering operator and the *free* evolution group, which, in the Schrödinger case, has a very simple form, contrarily to the total evolution group. This will prove to be very useful in the next section.

In the above we have written $g = F_+g + (I - F_+)g$ and seen that F_+g describes a part of g that will be outgoing as $t \to +\infty$. To interpret the remaining part $(I - F_+)g$, one has to assume asymptotic completeness in some sense. If one has generalized asymptotic completeness, then $(I - F_+)g$ will be an absorbed state at $t = +\infty$. In this case the picture is as follows : The scattering event is initiated in the state f (or rather $U_t^0 f$ at $t << -1$). As $t \to +\infty$, a part of the state will be absorbed, whereas another part, given by $U_t^0 Sf$, will propagate away to infinity. It is this latter part that is analyzed in a scattering experiment; hence it is this part that must be used to compute the scattering cross section (see the next section).

The situation becomes simpler if one assumes asymptotic completeness in the ordinary or in the geometric sense. One then has $R(\Omega_+) = R(\Omega_-)$, so that $g = \Omega_- f \in R(\Omega_+)$, i.e. $F_+g = g$. In this case there is no absorbed part (although the subspaces $\overline{M}_\Gamma^\pm(H)$ need not be empty).

If for instance $M_\infty^\pm(H_0) = H$, then in the latter situation, where $R(\Omega_+) = R(\Omega_-)$, S is unitary by Proposition 6.5, whereas in the former situation S is in general not unitary. It is thus seen that the unitarity or non-unitarity (more precisely the isometry or non-isometry) of S is concurrent with the impossibility or possibility respectively of absorbing at $t = +\infty$ a part of the scattering states at $t = -\infty$ (i.e. of the states in $M_\infty^-(H)$).

6.2 Scattering into Cones

In this section we present a mathematical model for a scattering experiment of the type mentioned at the beginning of this chapter. We assume for the time being that the target consists of a single scatterer, and we restrict ourselves to the case where the Hilbert space corresponding to a particle of the beam is $L^2(\mathbb{R}^\nu)$ and its free evolution is the Schrödinger free evolution introduced in Section 4.4. We also assume that $\nu \geq 2$.

We fix a direction $\underline{\omega}_o$, which will be the (approximate) direction of the velocity of the particles in the beam, and we denote by Π the $(\nu-1)$-dimensional hyperplane orthogonal to $\underline{\omega}_o$ that passes through the origin. The beam will be described by an ensemble of one-particle states, which are all identical except for translations by vectors in the plane Π. More precisely, we take the state of one of the particles of the beam to be $g \in L^2(\mathbb{R}^\nu)$, with $\|g\| = 1$, and the entire ensemble is the collection $\{g_{\underline{b}}\}_{\underline{b}\in\Pi}$, where $g_{\underline{b}}$ is the state obtained by translating g by the vector $\underline{b}$:

$$g_{\underline{b}}(\underline{x}) = g(\underline{x} - \underline{b}) \quad \text{or} \quad \tilde{g}_{\underline{b}}(\underline{k}) = e^{-i\underline{k}\cdot\underline{b}}\tilde{g}(\underline{k}). \tag{6.18}$$

In order to describe particles with relatively well defined momentum, we shall choose g such that $\tilde{g}$ has small support. For the moment we only assume that the support of $\tilde{g}$ is a compact subset of the half-space $\{\underline{k} \mid \underline{k}\cdot\underline{\omega}_o > 0\}$. We introduce the cone C_o spanned by the support of $\tilde{g}$, i.e. $C_o := \{\underline{k}\in\mathbb{R}^\nu \mid \alpha\underline{k}\in \operatorname{supp} \tilde{g} \text{ for some } \alpha > 0\}$, and we assume that $\underline{\omega}_o \in C_o$.

Let Δ be a Borel subset of $\mathbb{R}^\nu$ and $f\in L^2(\mathbb{R}^\nu)$, with $\|f\| = 1$. We denote by $P(f;\Delta)$ the probability that the scattering state $\Omega_- f$ corresponding to the initial state f will be localized in Δ (in configuration space) at $t = +\infty$, i.e.

$$P(f;\Delta) := \lim_{t\to+\infty} \int_\Delta |(U_t\Omega_- f)(\underline{x})|^2 d^\nu x. \tag{6.19}$$

If one adds the probabilities $P(g_{\underline{b}};\Delta)$ for all states of the ensemble $\{g_{\underline{b}}\}$, one gets a quantity representing the number $n(g;\Delta)$ of scattered particles that will be found asymptotically in Δ :

$$n(g;\Delta) = \int_\Pi P(g_{\underline{b}};\Delta) d^{\nu-1} b. \tag{6.20}$$

We are particularly interested in the case where Δ is a truncated cone $C(\rho) := \{\underline{x}\in C \,|\, |\underline{x}| > \rho\}$, where C is a cone with apex at the origin and ρ some non-negative number. The number $n(g;C(\rho))$ then corresponds to the number of particles that would be counted by an ideal detector spanning C and positioned at a sufficiently large distance $d \geq \rho$ from the scatterer, provided the experimental setup is such that the individual scattering events are independent of one another, so that in particular the scattered particles are uncorrelated.

The scattering cross section $\sigma(g;C)$ for scattering into the cone C is defined as the quotient of the number of particles scattered into C and the number of incoming particles per unit area of the plane Π. In our convention the latter number is just $\int_\Sigma d^{\nu-1} b$ over a set Σ of (planar) measure 1, i.e. it is equal to 1. Hence we have $\sigma(g;C) = n(g;C(\rho))$. We shall now show that under reasonable assumptions this quantity does not depend on ρ, provided that ρ is chosen large enough.

Lemma 6.6 : Assume that the wave operators $\Omega_\pm$ exist and that one of the two conditions (α) and (β) below is satisfied :

(α) $\mathcal{R}(\Omega_-) \subseteq \mathcal{R}(\Omega_+)$,

(β) one has generalized asymptotic completeness, i.e. (5.48), where

Γ is a closed set of measure zero contained in the interior of a ball B_{ρ_0} for some $\rho_0 \in (0,\infty)$.

Let $f \in L^2(\mathbb{R}^\nu)$, with $\|f\| = 1$. Then

$$P(f;C(\rho)) = \int_C |(\widetilde{Sf})(\underline{k})|^2 d^\nu k \tag{6.21}$$

for each $\rho \geq 0$ if one assumes (α) and for each $\rho \geq \rho_0$ if one assumes (β).

<u>Remark</u> : The result (6.21) is just what one expects intuitively : The probability of finding at $t = +\infty$ the scattering state associated with f in the truncated cone C is nothing but the probability that the outgoing part Sf of the corresponding final state has momentum in the cone C, i.e. the probability that Sf propagates in a direction along which it will ultimately have penetrated the cone C.

<u>Proof</u> : We denote by $F_\Delta = \chi_\Delta(\underline{Q})$ the operator that projects $L^2(\mathbb{R}^\nu)$ onto $L^2(\Delta)$. A special case is the operator $F_r \equiv F_{B_r}$ introduced in Section 5.1. We also set $F_+ = \Omega_+\Omega_+^*$ and we denote by F_Γ^+ the projection with range $\mathcal{M}_\Gamma^+(H)$.

(i) We may write

$$F_{C(\rho)}U_t\Omega_- f = F_{C(\rho)}U_t F_+\Omega_- f + F_{C(\rho)}U_t F_\Gamma^+\Omega_- f.$$

If $\mathcal{R}(\Omega_-) \subseteq \mathcal{R}(\Omega_+)$, the second term on the r.h.s. is zero. If (5.48) holds and $\rho \geq \rho_0$, the second term on the r.h.s. converges strongly to zero as $t \to +\infty$ by the definition of $\mathcal{M}_\Gamma^+(H)$. Hence, for $\rho \geq \rho_0$:

$$P(f;C(\rho)) = \lim_{t\to+\infty} \|F_{C(\rho)}U_t F_+\Omega_- f\|^2. \tag{6.22}$$

(ii) Using first the inequality $|a^2 - b^2 - c^2| \leq |a+b||a-b| + c^2$, then (1.15) and finally (1.24), we obtain

$$| \|F_{C(\rho)}U_tF_+\Omega_-f\|^2 - \|F_CU_t^oSf\|^2 |$$

$$= | \|F_{C(\rho)}U_tF_+\Omega_-f\|^2 - \|F_{C(\rho)}U_t^oSf\|^2 - \|F_CF_\rho U_t^oSf\|^2 |$$

$$\le 2| \|F_{C(\rho)}U_tF_+\Omega_-f\| - \|F_{C(\rho)}U_t^oSf\| | + \|F_CF_\rho U_t^oSf\|^2$$

$$\le 2\|F_{C(\rho)}[U_tF_+\Omega_-f - U_t^oSf]\| + \|F_CF_\rho U_t^oSf\|^2$$

$$\le 2\|F_+\Omega_-f - U_t^*U_t^o\Omega_+^*\Omega_-f\| + \|F_\rho U_t^oSf\|^2. \qquad (6.23)$$

As $t \to +\infty$, the first term on the r.h.s. converges to $2\|F_+\Omega_-f - \Omega_+\Omega_+^*\Omega_-f\| = 0$, and the second one converges to zero by Example 5.6. Consequently (6.22) and (6.23) imply that

$$P(f;C(\rho)) = \lim_{t\to+\infty} \|F_CU_t^oSf\|^2. \qquad (6.24)$$

(iii) In this last step we use Proposition 4.37. By (4.99) we have

$$\|F_CU_t^oSf\| = \|U_t^{o*}F_CU_t^oSf\| = \|Z_t^*\chi_C(2t\underline{P})Z_tSf\|$$

$$= \|\chi_C(2t\underline{P})Z_tSf\| = \|\chi_C(\underline{P})Z_tSf\|,$$

where the last equality holds because C is invariant under dilations, i.e. because $\underline{k}\in C \Leftrightarrow 2t\underline{k}\in C$ for $t > 0$, so that $\chi_C(2t\underline{k}) = \chi_C(\underline{k})$ if $t > 0$.

Now by (4.98), $Z_t \to I$ strongly as $t \to +\infty$, hence $\|\chi_C(\underline{P})Z_tSf\| \to \|\chi_C(\underline{P})Sf\|$ as $t \to +\infty$ by Proposition 1.1. Thus

$$P(f;C(\rho)) = \|\chi_C(\underline{P})Sf\|^2 = \int_C |\widetilde{Sf}(\underline{k})|^2 d^\nu k. \qquad \square$$

<u>Lemma 6.7</u> : Assume the hypotheses of Lemma 6.6. Let C_o be the cone spanned by supp $\tilde{g}$, and assume that C and C_o are disjoint, i.e. $C \cap C_o = \emptyset$. Then, for $\rho \ge \rho_o$:

$$P(g;C(\rho)) = \int_C |(\tilde{Rg})(\underline{k})|^2 d^\nu k. \tag{6.25}$$

Proof : Using (6.21) and $S = R + I$, we get

$$P(g;C(\rho)) = \int_C |(\tilde{Rg})(\underline{k})|^2 d^\nu k + \int_C |\tilde{g}(\underline{k})|^2 d^\nu k$$

$$+ 2\mathrm{Re} \int_C \overline{\tilde{g}(\underline{k})} (\tilde{Rg})(\underline{k}) d^\nu k.$$

Now the last two terms on the r.h.s. are zero, since supp $\tilde{g} \subset C_o$ and $C \cap C_o = \emptyset$ imply that $\tilde{g}(\underline{k}) = 0$ for $\underline{k} \in C$. □

To obtain an expression for $\sigma(g;C)$ lending itself to a practical interpretation, one must know that the R-matrix $R(\lambda) \equiv S(\lambda) - I_o$ is an integral operator in $L^2(S^{\nu-1})$ (i.e. with respect to the angles in the momentum variables). This is the case in particular when $R(\lambda)$ is a Hilbert-Schmidt operator (see Proposition 3.4). Hilbert-Schmidt properties of $R(\lambda)$ will be derived in Section 6.3 and will be assumed for the present discussion. In the next proposition we make a somewhat weaker hypothesis, viz. that the operator $R(\lambda)$ is Hilbert-Schmidt when sandwiched between two projections G_C and G_{C_o} the ranges of which are the states in $L^2(S^{\nu-1})$ with support in the intersection of $S^{\nu-1}$ with C and C_o respectively. Thus, for $h \in L^2(S^{\nu-1})$ and C a cone, we define

$$(G_C h)(\underline{\omega}) = \chi_{C \cap S^{\nu-1}}(\underline{\omega}) h(\underline{\omega}). \tag{6.26}$$

The kernel of the integral operator $G_C R(\lambda) G_{C_o}$ will be denoted by $r(\lambda;\underline{\omega},\underline{\omega}')$, where $\underline{\omega} \in C \cap S^{\nu-1}$ and $\underline{\omega}' \in C_o \cap S^{\nu-1}$. If $R(\lambda)$ is itself a Hilbert-Schmidt operator then $r(\lambda;\underline{\omega},\underline{\omega}')$ is defined for (almost) all $\underline{\omega},\underline{\omega}' \in S^{\nu-1}$. We also define the energy support $\Sigma(g)$ of a vector $g \in L^2(\mathbb{R}^\nu)$ to be the set $\Sigma(g) := \{\lambda \mid \lambda = \underline{k}^2$ for some $\underline{k} \in \text{supp } \tilde{g}\}$.

Proposition 6.8 : Assume the hypotheses of Lemma 6.6. Let $g \in L^2(\mathbb{R}^\nu)$

be such that $||g|| = 1$, $\tilde{g} \in L^{\infty}(\mathbb{R}^{\nu})$ and $\mathrm{supp}\,\tilde{g} \subseteq \{\underline{k} | \underline{k} \cdot \underline{\omega}_o > 0\}$. Let C_o be the cone spanned by supp $\tilde{g}$ and let $C \cap C_o = \emptyset$. Also assume that

$$\int_{\Sigma(g)} ||G_C R(\lambda) G_{C_o}||^2_{HS} d\lambda < \infty. \tag{6.27}$$

Then $\sigma(g;C)$ is finite and given by

$$\sigma(g;C) = \int_C d\lambda d\omega \int d\omega' \left(\frac{2\pi}{\sqrt{\lambda}}\right)^{\nu-1} |r(\lambda;\underline{\omega},\underline{\omega}')|^2 |(\mathcal{U}_o g)_{\lambda}(\underline{\omega}')|^2 [\cos(\underline{\omega}_o \cdot \underline{\omega}')]^{-1}. \tag{6.28}$$

Proof : We set $C_u := C \cap S^{\nu-1}$ and $C_{ou} := C_o \cap S^{\nu-1}$.

(i) (6.27) implies that $G_C R(\lambda) G_{C_o} \in \mathcal{B}_2(L^2(S^{\nu-1}))$ for almost all $\lambda \in \Sigma(g)$, so that $r(\lambda;\underline{\omega},\underline{\omega}')$ exists for almost all $\lambda \in \Sigma(g)$. Using successively (6.20), (6.25), (5.82), (3.11) and (6.18) and noticing that supp $\tilde{g}_{\underline{b}}$ = supp $\tilde{g}$, we get

$$\sigma(g;C) = \int_{\Pi} P(g_{\underline{b}};C(\rho)) d^{\nu-1}b = \int_{\Pi} d^{\nu-1}b \int_C d^{\nu}k |(\mathcal{F}Rg_{\underline{b}})(\underline{k})|^2$$

$$= \int_{\Pi} d^{\nu-1}b \int_{\Sigma(g)} d\lambda ||G_C R(\lambda) G_{C_o} (\mathcal{U}_o g_{\underline{b}})_{\lambda}||^2_o$$

$$= \int_{\Pi} d^{\nu-1}b \int_{\Sigma(g)} d\lambda \int_{C_u} d\omega | \int d\omega' r(\lambda;\underline{\omega},\underline{\omega}') (\mathcal{U}_o g_{\underline{b}})_{\lambda}(\underline{\omega}')|^2$$

$$= \int_{\Pi} d^{\nu-1} \int_{\Sigma(g)} d\lambda \int_{C_u} d\omega | \int d\omega' e^{-i\sqrt{\lambda}\underline{\omega}' \cdot \underline{b}} r(\lambda;\underline{\omega},\underline{\omega}') (\mathcal{U}_o g)_{\lambda}(\underline{\omega}')|^2.$$

Let us interchange the integration over $d^{\nu-1}b$ with those over $d\lambda$ and $d\omega$. The integral $J(\lambda,\underline{\omega}) := \int d^{\nu-1}b|\cdots|^2$ has essentially the form of the norm of the Fourier transform of a certain function, and can be reduced to that form by a suitable change of variables in the integral over $d\omega'$. One may therefore evaluate $J(\lambda,\underline{\omega})$ by using the unitarity of the Fourier transformation in $L^2(\mathbb{R}^{\nu-1})$, i.e. (1.18). This will be done in (ii) below. We leave it to the reader to check that insertion of the final expression (6.29) for $J(\lambda,\underline{\omega})$ into the equation

$$\sigma(g;C) = \int_{\Sigma(g)} d\lambda \int_{C_u} d\omega J(\lambda,\underline{\omega})$$

gives precisely the identity (6.28) that we set out to prove.

(ii) Let $\varphi_{\lambda,\underline{\omega}}(\underline{\omega}') = r(\lambda;\underline{\omega},\underline{\omega}')(U_o g)_\lambda(\underline{\omega}')\chi_{C_{ou}}(\underline{\omega}')$. We notice that $\varphi_{\lambda,\underline{\omega}}(\cdot)\in L^2(S^{\nu-1})$ for almost all λ and $\underline{\omega}$ by (6.27) and the assumption that $\tilde{g}\in L^\infty(\mathbb{R}^\nu)$. Also $\varphi_{\lambda,\underline{\omega}}(\cdot)$ has support on the half-sphere $\{\underline{\omega}'|\underline{\omega}_o\cdot\underline{\omega}' > 0\}$. We may therefore parametrize the points $\underline{\omega}'$ in the support of $\varphi_{\lambda,\underline{\omega}}$ by their orthogonal projection $\underline{z}$ onto Π; in other words we write $\underline{\omega}' = \underline{z} + (\underline{\omega}_o\cdot\underline{\omega}')\underline{\omega}_o$ with $\underline{z}\in\Pi$ and make the change of variables $\underline{\omega}' \mapsto \underline{z}$. We then have $d\omega' = [\cos(\underline{\omega}_o\cdot\underline{\omega}')]^{-1}d^{\nu-1}z$ and $\underline{\omega}'\cdot\underline{b} = \underline{z}\cdot\underline{b}$. Setting $\underline{b}' = \sqrt{\lambda}\underline{b}$ we obtain by using the unitarity of the Fourier transformation :

$$J(\lambda,\underline{\omega}) = \int\frac{d^{\nu-1}b'}{\lambda^{(\nu-1)/2}}\left|\int d^{\nu-1}z\; e^{-i\underline{z}\cdot\underline{b}'}\varphi_{\lambda,\underline{\omega}}(\underline{\omega}'(\underline{z}))[\cos(\underline{\omega}_o\cdot\underline{\omega}'(\underline{z}))]^{-1}\right|^2$$

$$= \left(\frac{2\pi}{\sqrt{\lambda}}\right)^{\nu-1}\int d^{\nu-1}z\,|\varphi_{\lambda,\underline{\omega}}(\underline{\omega}'(\underline{z}))|^2|\cos(\underline{\omega}_o\cdot\underline{\omega}'(\underline{z}))|^{-2}$$

$$= \left(\frac{2\pi}{\sqrt{\lambda}}\right)^{\nu-1}\int d\omega'\,|\varphi_{\lambda,\underline{\omega}}(\underline{\omega}')|^2|\cos(\underline{\omega}_o\cdot\underline{\omega}')|^{-1}. \qquad (6.29)$$

(iii) Since $\cos(\underline{\omega}_o\cdot\underline{\omega}') \geq \delta > 0$ for all $\underline{\omega}'\in C_{ou}$, we deduce from (6.28) and (5.83) that

$$\sigma(g;C) \leq (2\pi)^{\nu-1}\delta^{-1}\int_{\Sigma(g)} d\lambda\,\lambda^{(1-\nu)/2}\|G_C R(\lambda)G_{C_o}\|^2_{HS}\,\lambda^{(\nu-2)/2}\|\tilde{g}\|^2_\infty$$

$$\leq (2\pi)^{\nu-1}\delta^{-1}\|\tilde{g}\|^2_\infty[\inf\Sigma(g)]^{-1/2}\int_{\Sigma(g)} d\lambda\,\|G_C R(\lambda)G_{C_o}\|^2_{HS},$$

which is finite under the hypotheses of the proposition. □

We add a few comments on the interpretation of (6.28). The quantity $|(U_o g)_\lambda(\underline{\omega}')|^2$ is the probability $P(\lambda,\underline{\omega}')$ that an incoming particle has kinetic energy λ and direction of velocity $\underline{\omega}'$. As already said, this probability should be zero (or very small) except for $(\lambda,\underline{\omega}')$ in some small neighbourhood of some point $(\lambda_o,\underline{\omega}_o)$. Under this assumption $\cos(\underline{\omega}_o\cdot\underline{\omega}') \approx 1$, and this factor may be neglected in (6.28). One may then interpret

$$\sigma(\lambda,\underline{\omega}';C) := \left(\frac{2\pi}{\sqrt{\lambda}}\right)^{\nu-1} \int_C |r(\lambda;\underline{\omega},\underline{\omega}')|^2 d\omega \tag{6.30}$$

as the scattering cross section into the cone C for a beam consisting of particles of energy λ and direction of velocity $\underline{\omega}'$. Indeed (6.28) then just reads

$$\sigma(g;C) = \int P(\lambda,\underline{\omega}')\sigma(\lambda,\underline{\omega}';C)d\lambda d\omega'.$$

Another way of arriving at this interpretation is to take in (6.28) a sequence $\{g_n\}$ of state vectors such that $|(U_o g_n)_\lambda(\underline{\omega}')|^2$ converges to a delta-function $\delta(\lambda-\lambda_o)\delta(\underline{\omega}'-\underline{\omega}_o)$. Under suitable continuity assumptions on $r(\lambda;\underline{\omega},\underline{\omega}')$, one can then show that $\lim \sigma(g_n;C)$ as $n \to \infty$ is given by (6.30). We refer to [AJS; p. 287] for details.

In view of these remarks, one interprets the quantity $\sigma(\lambda,\underline{\omega}_o;C)$ as the scattering cross section for a beam consisting of particles with sharp momentum $\underline{k}_o = \sqrt{\lambda}\underline{\omega}_o$, and for any cone C such that $C \cap C_o = \emptyset$ for some open cone C_o with $\underline{\omega}_o \in C_o$. (6.30) shows that, for fixed λ and $\underline{\omega}_o$, $\sigma(\lambda,\underline{\omega}_o;C)$ defines an absolutely continuous measure on $S^{\nu-1}\backslash\{\omega_o\}$, with respect to Lebesgue measure. Its Radon-Nikodym derivative is called the <u>differential scattering cross section</u> $d\sigma/d\omega$:

$$\frac{d\sigma}{d\omega}(\lambda,\underline{\omega}_o;\underline{\omega}) := \left(\frac{2\pi}{\sqrt{\lambda}}\right)^{\nu-1}|r(\lambda;\underline{\omega},\underline{\omega}_o)|^2. \tag{6.31}$$

In addition to this, one considers the <u>total scattering cross section</u> σ_{tot}, defined as the integral of $d\sigma/d\omega$ over all final directions $\underline{\omega}$, and the <u>averaged total scattering</u> cross section $\bar{\sigma}$ defined as the average of σ_{tot} over all initial directions $\underline{\omega}_o$. Thus

$$\sigma_{tot}(\lambda,\underline{\omega}_o) := \sigma(\lambda,\underline{\omega}_o;S^{\nu-1}\setminus\{\underline{\omega}_o\}) = \left(\frac{2\pi}{\sqrt{\lambda}}\right)^{\nu-1} \int_{S^{\nu-1}} |r(\lambda;\underline{\omega},\underline{\omega}_o)|^2 d\omega \quad (6.32)$$

and

$$\bar{\sigma}(\lambda) := \frac{1}{\Theta_\nu} \int_{S^{\nu-1}} \sigma_{tot}(\lambda,\underline{\omega}_o)\, d\omega_o = \frac{1}{\Theta_\nu}\left(\frac{2\pi}{\sqrt{\lambda}}\right)^{\nu-1} \iint |r(\lambda;\underline{\omega},\underline{\omega}_o)|^2 d\omega d\omega_o$$

$$= \frac{1}{\Theta_\nu}\left(\frac{2\pi}{\sqrt{\lambda}}\right)^{\nu-1} \|R(\lambda)\|_{HS}^2 , \quad (6.33)$$

where Θ_ν denotes the surface area of $S^{\nu-1}$.

In particular we see that $\bar{\sigma}(\lambda)$ is simply related to the Hilbert-Schmidt norm of $R(\lambda)$ and that the finiteness of $\sigma_{tot}(\lambda,\underline{\omega}_o)$ for almost all $\underline{\omega}_o$ is implied by the condition that $R(\lambda)\in\mathcal{B}_2(L^2(S^{\nu-1}))$.

In the above we have assumed that the target consists of only one scatterer. For a target consisting of N *randomly distributed* scatterers, the various expressions for the scattering cross section given above have to be multiplied by N. More special situations will have to be treated differently.

6.3 Bounds on Scattering Cross Sections

We have seen that the total scattering cross section $\sigma_{tot}(\lambda,\underline{\omega}_o)$ is finite for almost all initial directions $\underline{\omega}_o$ if $R(\lambda) \equiv S(\lambda) - I_o$ is a Hilbert-Schmidt operator. In this last section we introduce an essentially time-dependent method for proving that $R(\lambda)$ is Hilbert-Schmidt for almost all λ. At the same time we obtain bounds on weighted averages of the scattering cross section over a range of energies, as well as conditions for the validity of the inequality (6.27) which was needed in our discussion of scattering into cones.

One of the difficulties in studying $R(\lambda)$ for fixed λ is that this operator does not act in the same Hilbert space as the operators that define the scattering theory, such as U_t, U_t^o, F_r etc. However, if one is satisfied with averages over the energy λ of quantities involving $R(\lambda)$, one can sometimes express such averages in terms of these latter operators. The method presented below is of this kind.

Proposition 6.9 : Let $\psi : [0,\infty) \to \mathbb{R}$ be a measurable function such that $||\psi||^2 := \int_0^\infty |\psi(\lambda)|^2 d\lambda = 1$. Let $P(\psi)$ be the following operator in $L^2(\mathbb{R}^\nu)$:

$$[\mathcal{U}_o P(\psi) f]_\lambda = \psi(\lambda) \int_0^\infty \psi(\mu)(\mathcal{U}_o f)_\mu d\mu. \tag{6.34}$$

Then (a) $P(\psi)$ is an orthogonal projection, and its range is the subspace

$$\mathcal{H}(\psi) := \{f \in L^2(\mathbb{R}^\nu) \mid (\mathcal{U}_o f)_\lambda = \psi(\lambda) g \text{ for some } g \in L^2(S^{\nu-1})\}. \tag{6.35}$$

(b) If the wave operators $\Omega_\pm$ exist, one has

$$\begin{aligned} ||RP(\psi)||_{HS}^2 &= \int_0^\infty |\psi(\lambda)|^2 ||R(\lambda)||_{HS}^2 d\lambda \\ &= (2\pi)^{1-\nu} \Theta_\nu \int_0^\infty \lambda^{(\nu-1)/2} |\psi(\lambda)|^2 \bar{\sigma}(\lambda) d\lambda. \end{aligned} \tag{6.36}$$

Comments : (i) The integral in (6.34) is a vector-valued integral in $L^2(S^{\nu-1})$. We have not given the general definition of such integrals. However, in later applications ψ will be a continuous function, and one may use the Riemann integral introduced in Section 1.3 to define $P(\psi)$ on the dense set of vectors f such that $\mu \mapsto (\mathcal{U}_o f)_\mu$ is strongly continuous.

(ii) Using (1.57) and the Schwarz inequality, it is easy to see that $P(\psi) \in \mathcal{B}(\mathcal{H})$:

$$||P(\psi)f||^2 = \int_0^\infty d\lambda\, \psi(\lambda)^2 \;||\int_0^\infty \psi(\mu)(\mathcal{U}_o f)_\mu d\mu||_0^2$$

$$\leq ||\psi||^2 [\int_0^\infty |\psi(\mu)|\, ||(\mathcal{U}_o f)_\mu||_0 d\mu]^2 \leq ||\psi||^4 \int_0^\infty ||(\mathcal{U}_o f)_\mu||_0^2 d\mu = ||f||^2.$$

(iii) (6.35) states that the range of $P(\psi)$ are those functions the Fourier transforms of which factorize into the product of a function of the angle $\underline{\omega} = \underline{k}/|\underline{k}|$ and a function of $|\underline{k}|$, viz. $|\underline{k}|^{(2-\nu)/2}\psi(|\underline{k}|^2)$.

(iv) (6.36) expresses an integral involving $R(\lambda)$ in terms of a Hilbert-Schmidt norm in $L^2(\mathbb{R}^\nu)$, as indicated before. The second identity in (6.36) follows immediately from (6.33).

Proof : (a) (i) Clearly $\mathcal{U}_o P(\psi)f$, defined by (6.34), has the form indicated in (6.35), i.e. $P(\psi)f \in \mathcal{H}(\psi)$. On the other hand, if $f \in \mathcal{H}(\psi)$, then

$$[\mathcal{U}_o P(\psi)f]_\lambda = \psi(\lambda) \int_0^\infty \psi(\mu)^2 g\, d\mu = \psi(\lambda)g = (\mathcal{U}_o f)_\lambda.$$

Hence $P(\psi)f = f$, i.e. $f \in \mathcal{R}(P(\psi))$. This shows that $\mathcal{H}(\psi) \subseteq \mathcal{R}(P(\psi))$. Hence $\mathcal{R}(P(\psi)) = \mathcal{H}(\psi)$.

(ii) A calculation similar to that above leads to

$$[\mathcal{U}_o P(\psi)^2 f]_\lambda = \psi(\lambda) \int_0^\infty \psi(\mu)[\mathcal{U}_o P(\psi)f]_\mu d\mu$$

$$= \psi(\lambda) \int_0^\infty \psi(\mu)^2 d\mu \int_0^\infty \psi(\mu')(\mathcal{U}_o f)_{\mu'} d\mu'$$

$$= \psi(\lambda) \int_0^\infty \psi(\mu')(\mathcal{U}_o f)_{\mu'} d\mu' = [\mathcal{U}_o P(\psi)f]_\lambda.$$

Hence $P(\psi)^2 f = P(\psi)f$, showing that $P(\psi)$ is idempotent.

(iii) Finally, using the hypothesis that ψ is real-valued and interchanging the order of integration, we find that

$$(g,P(\psi)f) = \int_0^\infty d\lambda\, \psi(\lambda) \int_0^\infty d\mu\, \psi(\mu)((\mathcal{U}_o g)_\lambda, (\mathcal{U}_o f)_\mu)_0 = (P(\psi)g,f).$$

Hence $P(\psi)^* = P(\psi)$, which proves that $P(\psi)$ is a projection.

(b) (i) Let $\{e_k\}$ be an orthonormal basis of $L^2(S^{\nu-1})$, and define $h_k \in H(\psi)$ by $(U_o h_k)_\lambda = \psi(\lambda)e_k$. Then the vectors $\{h_k\}$ form an orthonormal basis of $H(\psi)$. Indeed $(h_k, h_m) = (e_k, e_m)_o = \delta_{km}$, and if $f \in H(\psi)$, then by (5.82)

$$(f, h_k) = \int_0^\infty (\psi(\lambda)g, \psi(\lambda)e_k)_o d\lambda = (g, e_k)_o ,$$

and

$$\sum_k |(g, e_k)_o|^2 = ||g||_o^2 = \int_0^\infty ||\psi(\lambda)g||_o^2 d\lambda = ||f||^2 .$$

(ii) From the definition (3.1) of the Hilbert-Schmidt norm, we have $||RP(\psi)||_{HS}^2 = \sum_k ||RP(\psi)g_k||^2$, where $\{g_k\}$ is any orthonormal basis of $L^2(\mathbb{R}^\nu)$. Since $P(\psi)$ is a projection, it suffices to take for $\{g_k\}$ an orthonormal basis of $H(\psi)$, for instance the set $\{h_k\}$ defined above. Then

$$||RP(\psi)||_{HS}^2 = \sum_k ||Rh_k||^2 = \sum_k \int_0^\infty ||R(\lambda)[\psi(\lambda)e_k]||_o^2 d\lambda$$

$$= \sum_k \int_0^\infty \psi(\lambda)^2 ||R(\lambda)e_k||_o^2 d\lambda = \int_0^\infty \psi(\lambda)^2 ||R(\lambda)||_{HS}^2 d\lambda ,$$

where we have used the following facts : (α) R is decomposable, and (β) for fixed λ, $\psi(\lambda)$ is just a number. □

Corollary 6.10 : Suppose that the wave operators $\Omega_\pm$ exist. If $RP(\psi)$ is a Hilbert-Schmidt operator for some ψ, then $R(\lambda)$ is Hilbert-Schmidt for almost all λ belonging to $\Delta_\psi := \{\lambda | \psi(\lambda) \neq 0\}$. If the hypotheses of Lemma 6.6 are satisfied, then $\sigma_{tot}(\lambda, \underline{\omega})$ is finite for almost all $\lambda \in \Delta_\psi$ and $\underline{\omega} \in S^{\nu-1}$.

It remains to find conditions on the function ψ and the potential v ensuring that $RP(\psi)\in\mathcal{B}_2$. We first explain formally how such conditions may be obtained. We assume that $S = W_+^* W_-$ with $W_\pm = \text{s-lim}\ U_t^* J U_t^0$ as $t \to \pm\infty$, where $J\in\mathcal{B}(\mathcal{H})$ is independent of the sign of time, and $\text{s-lim}(I - J)U_t^o = 0$ as $t \to \pm\infty$, so that $W_\pm$ are isometries. Then

$$RP(\psi) = (S - I)P(\psi) = W_+^*(W_- - W_+)P(\psi)$$

$$= -W_+^* \int_{-\infty}^{\infty} \frac{d}{dt} U_t^* J U_t^0 P(\psi)\,dt = -iW_+^* \int_{-\infty}^{\infty} U_t^*(HJ - JH_0)U_t^0 P(\psi)\,dt.$$

The derivative exists if $\mathcal{H}(\psi) \subseteq D(H_0)$ and $JD(H_0) \subset D(H)$. Using the inequality (1.57) in the Hilbert space $\mathcal{B}_2$ and (3.5), we arrive at

$$\|RP(\psi)\|_{HS} \le \int_{-\infty}^{\infty} \|W_+^* U_t^*\|\, \|(HJ - JH_0)U_t^0 P(\psi)\|_{HS}\,dt$$

$$= \int_{-\infty}^{\infty} \|(HJ - JH_0)U_t^0 P(\psi)\|_{HS}\,dt. \tag{6.37}$$

In concrete situations the operators appearing under the integral sign are explicitly given, so that it should be easy to estimate the integral. This is what we shall do below.

Suppose that the integral in (6.37) is finite and that $\mathcal{H}(\psi) \subseteq D(H_0)$ (i.e. that $\lambda\psi(\lambda)\in L^2(\mathbb{R})$). This has the following consequences : (a) For each $f\in\mathcal{H}(\psi)$, we have from (5.54) that

$$\int_{-\infty}^{\infty} \left\|\frac{d}{dt} U_t^* J U_t^0 f\right\| dt < \infty. \tag{6.38}$$

Hence the wave operators $W_\pm$ exist on the subspace $\mathcal{H}(\psi)$, see Lemma 5.18. If the integral in (6.37) is finite for all ψ in some class A of functions and if the set $\cup_{\psi\in A}\mathcal{H}(\psi)$ is total in $\mathcal{H}$, then Proposition 5.19 implies the existence of $W_\pm = \text{s-lim}\ U_t^* J U_t^0$. By Proposition 5.17, the wave operators $\Omega_\pm = \text{s-lim}\ U_t^* U_t^0$ as $t \to \pm\infty$ will then also exist. (b) $RP(\psi)\in\mathcal{B}_2$, hence $\bar{\sigma}(\lambda) < \infty$ for almost all $\lambda\in\Delta_\psi$.

(c) The condition (6.27) is satisfied if $\Sigma(g) \subseteq \{\lambda \mid |\psi(\lambda)| > \varepsilon > 0\}$.

In the next two lemmas we give some preliminary results that will allow us to estimate integrals of the type (6.37). Afterwards we shall apply these results to discuss the following three questions : finiteness of the scattering cross section, high energy behaviour of the cross section, bounds on the cross section for potentials of finite range.

The following notation will be used : if $\alpha \in \mathbb{R}$, then we denote by ψ_α the function $\psi_\alpha(\lambda) := \lambda^{\alpha/4}\psi(\lambda)$ and by ψ'_α its derivative.

Lemma 6.11 : Let $w \in L^2(\mathbb{R}^\nu)$ be real-valued and denote by W the self-adjoint multiplication operator by $w(\underline{x})$. Also denote by Θ_ν the surface area of $S^{\nu-1}$. Let $\psi : (0,\infty) \to \mathbb{R}$ be continuously differentiable, $\psi = 0$ in a neighbourhood of 0 and of ∞ and $\|\psi\|^2 = 1$. Then, if U_t^o is the Schrödinger free evolution :

$$\int_{-\infty}^{\infty} \|WU_t^oP(\psi)\|_{HS}^2 dt = \frac{1}{2}(2\pi)^{1-\nu}\Theta_\nu \|\psi_{\nu-2}\|^2 \|w\|^2, \tag{6.39}$$

$$\int_{-\infty}^{\infty} \|WP_mU_t^oP(\psi)\|_{HS}^2 dt = \frac{1}{2\nu}(2\pi)^{1-\nu}\Theta_\nu \|\psi_\nu\|^2 \|w\|^2, \tag{6.40}$$

$$\begin{aligned}\int_{-\infty}^{\infty} t^2 \|(I+|\underline{Q}|)^{-1}WU_t^oP(\psi)\|_{HS}^2 dt &= \frac{1}{2}(2\pi)^{1-\nu}\Theta_\nu \|\psi'_{\nu-2}\|^2 \|(I+|\underline{Q}|)^{-1}w\|^2 \\ &+ \frac{1}{8\nu}(2\pi)^{1-\nu}\Theta_\nu \|\psi_{\nu-4}\|^2 \|\,|\underline{Q}|(I+|\underline{Q}|)^{-1}w\|^2,\end{aligned} \tag{6.41}$$

$$\begin{aligned}\int_{-\infty}^{\infty} t^2 \|(I+|\underline{Q}|)^{-1}WP_mU_t^oP(\psi)\|_{HS}^2 &= \frac{1}{2\nu}(2\pi)^{1-\nu}\Theta_\nu \|\psi'_\nu\|^2 \|(I+|\underline{Q}|)^{-1}w\|^2 \\ &+ \frac{1}{8\nu(\nu+2)}(2\pi)^{1-\nu}\Theta_\nu \|\psi_{\nu-2}\|^2 [\|\,|\underline{Q}|(I+|\underline{Q}|)^{-1}w\|^2 \\ &+ 2\|Q_m(I+|\underline{Q}|)^{-1}w\|^2].\end{aligned} \tag{6.42}$$

Remark : It is interesting to notice that the above expressions are exact and that each term on the right-hand side is a product of a factor depending only on ψ and a factor depending only on w.

Proof : (i) Taking adjoints, we have $||WU_t^oP(\psi)||_{HS} = ||P(\psi)U_{-t}^oW||_{HS}$. To calculate this Hilbert-Schmidt norm, we write $P(\psi)U_{-t}^oW$ as an integral operator in momentum space and use Proposition 3.4. By interchanging in (3.19) the roles of $\underline{P}$ and $\underline{Q}$, we see that W is an integral operator in $\tilde{L}^2(\mathbb{R}^\nu)$ with kernel $(2\pi)^{-\nu/2}\tilde{w}(\underline{k}-\underline{k}')$. Using (6.34) and (5.83), we see that the kernel $N_t(\underline{k},\underline{k}')$ of $P(\psi)U_{-t}^oW$ is given by

$$N_t(\underline{k},\underline{k}') = (2\pi)^{-\nu/2}\lambda^{(2-\nu)/4}\psi(\lambda)\int_0^\infty d\mu\, \mu^{(\nu-2)/4}\psi(\mu)\tilde{w}(\sqrt{\mu}\underline{\omega}-\underline{k}')e^{i\mu t}, \tag{6.43}$$

where $\underline{k} = \lambda^{1/2}\underline{\omega}$. Noticing that $d^\nu k = \frac{1}{2}\lambda^{(\nu-2)/2}d\lambda d\omega$, we obtain from Proposition 3.4 that

$$||WU_t^oP(\psi)||_{HS}^2 = \int\int d^\nu k d^\nu k' |N_t(\underline{k},\underline{k}')|^2$$

$$= \frac{1}{2}(2\pi)^{-\nu}||\psi||^2\int d\omega d^\nu k' \left|\int_0^\infty d\mu\, \psi_{\nu-2}(\mu)\tilde{w}(\sqrt{\mu}\underline{\omega}-\underline{k}')e^{i\mu t}\right|^2. \tag{6.44}$$

The integration over $d\mu$ may be viewed as a one-dimensional Fourier transformation (up to a factor $(2\pi)^{1/2}$). This allows us to calculate the integral of (6.44) over the variable t by using the unitarity relation (1.18) for the Fourier transformation :

$$\int_{-\infty}^\infty ||WU_t^oP(\psi)||_{HS}^2 dt = \frac{1}{2}(2\pi)^{-\nu+1}\int d\omega d^\nu k' d\mu |\psi_{\nu-2}(\mu)\tilde{w}(\sqrt{\mu}\underline{\omega}-\underline{k}')|^2$$

$$= \frac{1}{2}(2\pi)^{1-\nu}\int d\omega d\mu |\psi_{\nu-2}(\mu)|^2||\tilde{w}||^2,$$

which proves (6.39).

(ii) The kernel of $P(\psi)U_{-t}^oP_mW$ is obtained from $N_t(\underline{k},\underline{k}')$ by inserting an additional factor $\mu^{1/2}\underline{\omega}\cdot\underline{\varepsilon}_m$, where $\underline{\varepsilon}_m$ is a unit vector along the m-th coordinate axis. The integral in (6.40) may be calculated as above. Instead of $||\lambda^{(\nu-2)/4}\psi||^2$ we now get $||\lambda^{\nu/4}\psi||^2$, because of the additional factor $\mu^{1/2}$ in the kernel. Also Θ_ν must be replaced by $\int|\underline{\omega}\cdot\underline{\varepsilon}_m|^2 d\omega$. This integral is easy to calculate by noticing that its value is independent of m. Since $\underline{\omega} = \sum_m(\underline{\omega}\cdot\underline{\varepsilon}_m)\underline{\varepsilon}_m$, we have $\underline{\omega}^2 = 1 = \sum_m|\underline{\omega}\cdot\underline{\varepsilon}_m|^2$, hence

$$\Theta_\nu = \sum_m\int|\underline{\omega}\cdot\underline{\varepsilon}_m|^2 d\omega = \nu\int|\underline{\omega}\cdot\underline{\varepsilon}_m|^2 d\omega \text{ for any } m = 1,\ldots,\nu.$$

This shows that

$$\int|\underline{\omega}\cdot\underline{\varepsilon}_m|^2 d\omega = \Theta_\nu/\nu, \tag{6.45}$$

which proves (6.40).

For later reference we also write down the values of the following integrals :

$$\int\underline{\omega}\cdot\underline{\varepsilon}_m d\omega = 0, \tag{6.46}$$

$$\int(\underline{\omega}\cdot\underline{\varepsilon}_m)(\underline{\omega}\cdot\underline{\varepsilon}_n)d\omega = 0 \quad \text{if} \quad m \neq n, \tag{6.47}$$

$$\int(\underline{\omega}\cdot\underline{\varepsilon}_m)^2(\underline{\omega}\cdot\underline{\varepsilon}_n)^2 d\omega = \frac{\Theta_\nu}{\nu(\nu+2)}(1 + 2\delta_{mn}), \tag{6.48}$$

$$\int(\underline{\omega}\cdot\underline{\varepsilon}_m)^2(\underline{\omega}\cdot\underline{\varepsilon}_n)d\omega = \int(\underline{\omega}\cdot\underline{\varepsilon}_m)^2(\underline{\omega}\cdot\underline{\varepsilon}_n)(\underline{\omega}\cdot\underline{\varepsilon}_k)d\omega = 0 \text{ if } k \neq n. \tag{6.49}$$

(iii) To prove (6.41), we set $w_0(\underline{x}) = (1 + |\underline{x}|)^{-1}w(\underline{x})$ and $w_m(\underline{x}) = x_m w_0(\underline{x})$, where $m = 1,\ldots,\nu$. The kernel of $tP(\psi)U^o_{-t}W(I + |\underline{Q}|)^{-1}$ is then given by

$$-i(2\pi)^{-\nu/2}\lambda^{(2-\nu)/4}\psi(\lambda)\int_0^\infty d\mu\, \mu^{(\nu-2)/4}\psi(\mu)\tilde{w}_0(\sqrt{\mu}\underline{\omega} - \underline{k}')\frac{d}{d\mu}e^{i\mu t}. \tag{6.50}$$

We first integrate by parts with respect to the variable μ and notice that the boundary terms vanish because $\psi(0) = \psi(\infty) = 0$. As in (i), we may then use the unitarity of the one-dimensional Fourier transformation to get that

$$\int_{-\infty}^{\infty} t^2 \| (I + |\underset{\sim}{Q}|)^{-1} W U_t^o P(\psi) \|_{HS}^2 = \frac{1}{2}(2\pi)^{1-\nu} \int d\omega d^\nu k' \int_0^\infty d\mu \cdot$$

$$\cdot | \tilde{w}_o(\sqrt{\mu}\underline{\omega} - \underline{k}')\psi'_{\nu-2}(\mu) + \psi_{\nu-2}(\mu) \frac{d}{d\mu}\tilde{w}_o(\sqrt{\mu}\underline{\omega} - \underline{k}') |^2. \qquad (6.51)$$

Now, setting $\omega_r = \underline{\omega} \cdot \underline{\varepsilon}_r$:

$$\frac{d}{d\mu}\tilde{w}_o(\sqrt{\mu}\underline{\omega} - \underline{k}') = \frac{1}{2}\mu^{-1/2} \frac{d}{d\sqrt{\mu}}\tilde{w}_o(\sqrt{\mu}\underline{\omega} - \underline{k}')$$

$$= \frac{1}{2}\mu^{-1/2} \frac{d}{d\sqrt{\mu}} (2\pi)^{-\nu/2} \int e^{-i\sqrt{\mu}\underline{\omega}\cdot\underline{x}} e^{i\underline{k}'\cdot\underline{x}} w_o(\underline{x}) d^\nu x$$

$$= -\frac{i}{2}\mu^{-1/2} \sum_{r=1}^{\nu} \omega_r \tilde{w}_r(\sqrt{\mu}\underline{\omega} - \underline{k}').$$

We insert this into (6.51), make the change of variables $\underline{k}' \mapsto \underline{k}'' = \sqrt{\mu}\underline{\omega} - \underline{k}'$ and obtain

$$\int_{-\infty}^{\infty} t^2 \| (I + |\underset{\sim}{Q}|)^{-1} W U_t^o P(\psi) \|_{HS}^2 =$$

$$= \frac{1}{2}(2\pi)^{1-\nu} \int d\mu d\omega \| \psi'_{\nu-2}(\mu)\tilde{w}_o(\cdot) - \frac{i}{2}\psi_{\nu-4}(\mu) \sum_{r=1}^{\nu} \omega_r \tilde{w}_r(\cdot) \|^2, \qquad (6.52)$$

where the norm is in $\tilde{L}^2(\mathbb{R}^\nu)$ w.r. to the variable $\underline{k}''$. The integral over $d\omega$ can now easily be evaluated by writing the norm as a sum of scalar products and using (6.45)-(6.47). One finds that (6.52) is equal to

$$\frac{1}{2}(2\pi)^{1-\nu}\Theta_\nu [\|\psi'_{\nu-2}\|^2 \|w_o\|^2 + \nu^{-1} \|\frac{1}{2}\psi_{\nu-4}\|^2 \sum_{r=1}^{\nu} \|w_r\|^2],$$

which is identical with the r.h.s. of (6.41).

(iv) The equation (6.42) can be derived by using the arguments of (ii) and (iii) above and the identities (6.45), (6.48) and (6.49). We omit the details. □

Problem 6.12 : In part (iii) of the preceding proof we have treated the derivatives with respect to μ a little carelessly. Our derivation is certainly justified if w is replaced by $w\chi_r$, where χ_r is the characteristic function of the ball B_r, since $Fw\chi_r$ is infinitely differentiable. Use this fact and a limiting argument to prove (6.41) and (6.42) for general w (e.g. show that $\|(I+|\underset{\sim}{Q}|)^{-1}WF_r U_t^o P(\psi)\|_{HS} \to \|(I+|\underset{\sim}{Q}|)^{-1}WU_t^o P(\psi)\|_{HS}$ as $r \to \infty$ and use the monotone convergence theorem).

Lemma 6.13 : Let w and ψ be as in Lemma 6.11. Let $\alpha \in (0,1)$, $\beta \geq 0$ and $\gamma > 0$. Then there is a constant $C_{\alpha\beta} \in (0,\infty)$, depending only on α and β, such that

$$\int_{-\infty}^{\infty} (1+\gamma^2 t^2)^{\alpha}[\log(2+\gamma|t|)]^{\beta}\|(I+|\underset{\sim}{Q}|)^{-\alpha}[\log(2+|\underset{\sim}{Q}|)]^{-\beta/2}WU_t^o P(\psi)\|_{HS}^2 dt$$
$$\leq \frac{1}{8}(2\pi)^{1-\nu}\Theta_{\nu}\|w\|^2[4(1+C_{\alpha\beta})\|\psi_{\nu-2}\|^2 + 4\gamma^2 C_{\alpha\beta}\|\psi'_{\nu-2}\|^2 +$$
$$+ \gamma^2 C_{\alpha\beta}\nu^{-1}\|\psi_{\nu-4}\|^2]. \qquad (6.53)$$

Proof : Denoting by $M_t(\underset{\sim}{x},\underset{\sim}{y})$ the kernel of the Hilbert-Schmidt operator $WU_t^o P(\psi)$, we see that the l.h.s. of (6.53) is equal to

$$\int_{-\infty}^{\infty} dt \int d^{\nu}x d^{\nu}y \frac{(1+\gamma^2 t^2)^{\alpha}}{(1+|\underset{\sim}{x}|)^{2\alpha}} \frac{[\log(2+\gamma|t|)]^{\beta}}{[\log(2+|\underset{\sim}{x}|)]^{\beta}} |M_t(\underset{\sim}{x},\underset{\sim}{y})|^2. \qquad (6.54)$$

We divide the domain of integration $\mathbb{R}^{2\nu+1}$ into two parts D_1 and D_2, where $D_1 = \{(t,\underset{\sim}{x},\underset{\sim}{y}) \mid \gamma|t| \leq |\underset{\sim}{x}|\}$ and $D_2 = \{(t,\underset{\sim}{x},\underset{\sim}{y}) \mid \gamma|t| > |\underset{\sim}{x}|\}$. In D_1 the integrand is majorized by $|M_t(\underset{\sim}{x},\underset{\sim}{y})|^2$. On the other hand in D_2 it is majorized by $C_{\alpha\beta}(1+\gamma^2t^2)(1+|\underset{\sim}{x}|)^{-2}|M_t(\underset{\sim}{x},\underset{\sim}{y})|^2$, since for $\alpha < 1$ and $\gamma|t| > |\underset{\sim}{x}|$:

$$[\log(2+\gamma|t|)/\log(2+|\underset{\sim}{x}|)]^{\beta} \leq C_{\alpha\beta}[(1+\gamma^2t^2)/(1+|\underset{\sim}{x}|)^2]^{1-\alpha}.$$

Notice that

$$C_{\alpha\beta} = \sup_{0\leq\rho<\tau<\infty} \frac{[\log(2+\tau)]^{\beta}}{[\log(2+\rho)]^{\beta}} \cdot \frac{(1+\rho)^{2-2\alpha}}{(1+\tau^2)^{1-\alpha}} < \infty.$$

By using the preceding bounds for the integrand in (6.54) and then extending the domains of integration D_1 and D_2 again to $\mathbb{R}^{2\nu+1}$, we see that (6.54) is majorized by

$$\int_{-\infty}^{\infty} dt[\|WU_t^oP(\psi)\|_{HS}^2 + C_{\alpha\beta}(1+\gamma^2t^2)\|(I+|\underline{Q}|)^{-1}WU_t^oP(\psi)\|_{HS}^2].$$

This integral may be evaluated by using (6.39) and (6.41). On the r.h.s. of (6.53) we have not given the exact expression but rather an upper bound which has a somewhat simpler form, by observing that all norms involving w on the r.h.s. of (6.39) and (6.41) are less than or equal to $\|w\|^2$. □

Proposition 6.14 : Let $v : \mathbb{R}^\nu \to \mathbb{R}$ be a potential. Define w by $w(\underline{x}) := (1+|\underline{x}|)^{1/2}[\log(2+|\underline{x}|)]^{(1+\varepsilon)/2}v(\underline{x})$, where $\varepsilon > 0$, and assume that $w \in L^2(\mathbb{R}^\nu)$. Let H be a self-adjoint extension of the operator $\hat{H} = -\Delta + v(\underline{x})$ defined on $D(\hat{H}) = S(\mathbb{R}^\nu)$, and let $\{U_t\}$ be the associated evolution group. Then the wave operators $\Omega_\pm =$ $=$ s-lim $U_t^*U_t^o$ as $t \to \pm\infty$ exist and the averaged total scattering cross section $\bar{\sigma}(\lambda)$ is finite for almost all $\lambda > 0$. Furthermore, for each real-valued ψ satisfying the conditions stated in Lemma 6.11, one has the following bound on $\bar{\sigma}(\lambda)$:

$$\int_0^\infty \lambda^{(\nu-1)/2}|\psi(\lambda)|^2\bar{\sigma}(\lambda)d\lambda \le C_\varepsilon d_\varepsilon \|(I+|\underline{Q}|)^{1/2}[\log(2+|\underline{Q}|)]^{(1+\varepsilon)/2}v\|^2 \cdot$$

$$\cdot[(4+\nu^{-1})\|\psi_{\nu-2}\|\,\|\psi_{\nu-4}\| + 4\|\psi'_{\nu-2}\|^2\|\psi_{\nu-2}\|\,\|\psi_{\nu-4}\|^{-1}], \qquad (6.55)$$

where $C_\varepsilon = \int_{-\infty}^{\infty}(1+t^2)^{-1/2}[\log(2+|t|)]^{-1-\varepsilon}dt$ and $d_\varepsilon = \frac{1}{8}(1+C_{\frac{1}{2},1+\varepsilon})$.

Proof : (i) We use (6.36) and (6.37), then multiply and divide the integrand by $(1+\gamma^2t^2)^{1/4}[\log(2+\gamma|t|)]^{(1+\varepsilon)/2}$ and apply the Schwarz inequality to obtain that for any $\gamma > 0$:

$$\int_0^\infty \lambda^{(\nu-1)/2}|\psi(\lambda)|^2\bar{\sigma}(\lambda)d\lambda \leq (2\pi)^{\nu-1}\Theta_\nu^{-1}[\int_{-\infty}^\infty \|VU_t^oP(\psi)\|_{HS}dt]^2$$

$$\leq (2\pi)^{\nu-1}\Theta_\nu^{-1}\int_{-\infty}^\infty (1+\gamma^2 s^2)^{-1/2}[\log(2+\gamma|s|)]^{-1-\varepsilon}ds\cdot$$

$$\cdot\int_{-\infty}^\infty (1+\gamma^2t^2)^{1/2}[\log(2+\gamma|t|)]^{1+\varepsilon}\|VU_t^oP(\psi)\|_{HS}^2dt$$

$$= (2\pi)^{\nu-1}\Theta_\nu^{-1}\gamma^{-1}C_\varepsilon \int_{-\infty}^\infty dt(1+\gamma^2t^2)^{1/2}[\log(2+\gamma|t|)]^{1+\varepsilon}\cdot$$

$$\cdot\|(I+|\underline{Q}|)^{-1/2}[\log(2+|\underline{Q}|)]^{-(1+\varepsilon)/2}WU_t^oP(\psi)\|_{HS}^2.$$

Inserting the bound (6.53) for the last integral and setting $\gamma = \|\psi_{\nu-2}\|\cdot\|\psi_{\nu-4}\|^{-1}$, one arrives at (6.55).

(ii) The finiteness a.e. of $\bar{\sigma}(\lambda)$ and the existence of $\Omega_\pm$ are consequences of Corollary 6.10 and of the discussion following it, by taking J = I. It is easy to see that the set $\{UH(\varphi)|\varphi\in C_{oo}^\infty(\mathbb{R})\}$ is total in $L^2(\mathbb{R}^\nu)$: If $g\in L^2(\mathbb{R}^\nu)$, set $G_k(\lambda) = ((\mathcal{U}_o g)_\lambda, e_k)_o$, where $\{e_k\}$ is an orthonormal basis of $L^2(S^{\nu-1})$. If $g \perp H(\psi)$ for all $\psi\in C_{oo}^\infty(\mathbb{R})$, then $\int G_k(\lambda)\psi(\lambda)d\lambda = 0$ $\forall k$ and $\forall\psi\in C_{oo}^\infty(\mathbb{R})$, which in turn implies that $G_k(\lambda) = \theta$ as a vector in $L^2(0,\infty)$, by Lemmas A.3 and 1.5. Hence $g = \theta$. □

In the corollary below we point out a condition on the decay of the potential v at infinity which ensures the finiteness of the total scattering cross section.

Corollary 6.15 : Assume that $v\in L^2_{loc}(\mathbb{R}^\nu)$ and that $|v(\underline{x})| \leq c(1+|\underline{x}|)^{-(\nu+1)/2}[\log(2+|\underline{x}|)]^{-1-\varepsilon}$ for $|\underline{x}| \geq a$, with $\varepsilon > 0$. Then $\bar{\sigma}(\lambda) < \infty$ a.e.

One can use (6.55) to obtain a good estimate of the high energy behaviour of the scattering cross section. For this, choose a continuously differentiable function $\varphi : \mathbb{R} \to \mathbb{R}$ having support in an interval (0,a) such that $\varphi(p) = 1$ on some subinterval [b,c] of $(0,a)$ and such that $\int|\varphi(p)|^2 dp = 1$, and take ψ of the form $\psi(\lambda) = (4\lambda)^{-1/4}\varphi(\sqrt{\lambda} - k_o)$, where $k_o > \max(1,a)$. It is easy to check that $\|\psi\| = 1$ and to obtain the following estimates :

$$\|\psi_{\nu-2}\|^2 \le c_\nu k_o^{\nu-2} \tag{6.56}$$

$$c'_\nu k_o^{\nu-4} \le \|\psi_{\nu-4}\|^2 \le c_\nu k_o^{\nu-4} \tag{6.57}$$

$$\|\psi'_{\nu-2}\|^2 \le \delta_\nu(\varphi) k_o^{\nu-4}, \tag{6.58}$$

where c_ν and c'_ν are positive constants depending on ν and $\delta_\nu(\varphi)$ is a constant depending on ν and φ. For example we have, setting $k = \sqrt{\lambda}$:

$$\|\lambda^{\alpha/4}\psi\|^2 = \frac{1}{2}\int \lambda^{(\alpha-1)/2}|\varphi(\sqrt{\lambda} - k_o)|^2 d\lambda$$

$$= \int k^\alpha |\varphi(k - k_o)|^2 dk = \int_0^a (p + k_o)^\alpha |\varphi(p)|^2 dp.$$

Using the bound $(p + k_o)^\alpha \le (2k_o)^\alpha$ if $\alpha \ge 0$ (since $p \le a \le k_o$) and $(p + k_o)^\alpha \le k_o^\alpha$ if $\alpha < 0$, one gets (6.56) and the second inequality in (6.57).

If we now take ψ as above in (6.55), we obtain from (6.56)-(6.58) that

$$\int_{(k_o+b)^2}^{(k_o+c)^2} \lambda^{(\nu-2)/2}\bar{\sigma}(\lambda)d\lambda \le 2\int_0^\infty \lambda^{(\nu-1)/2}|\psi(\lambda)|^2\bar{\sigma}(\lambda)d\lambda$$

$$\le \kappa_{\varepsilon\nu}(\varphi)\|(I + |Q|)^{1/2}[\log(2 + |Q|)]^{(1+\varepsilon)/2} V\|^2 k_o^{\nu-3}, \tag{6.59}$$

where $\kappa_{\varepsilon\nu}(\varphi)$ is a constant depending on ε,ν and φ. If we assume that $\bar{\sigma}(\lambda) = O(\lambda^{-\beta})$ as $\lambda \to \infty$, the first integral in (6.59) is $O(k_o^{\nu-2\beta-1})$. Hence (6.59) requires that $\beta \ge 1$. If we assume that $\bar{\sigma}(\lambda) = \Phi(\lambda)\lambda^{-1}$ with $\Phi(\lambda)$ non-decreasing, then the first integral in (6.59) is bounded below by a positive multiple of $\Phi(k_o^2)k_o^{\nu-3}$ as

$k_o \to \infty$. In this case (6.59) requires that Φ must be a *bounded* function.

Of course our bounds do not prove that $\bar{\sigma}(\lambda)$ converges to zero as $\lambda \to \infty$, since we consider only averages of $\bar{\sigma}(\lambda)$ over some interval. But if we assume the convergence to zero (or if it has been proven by other methods), then our bounds imply that the rate of convergence is at least as rapid as λ^{-1} (in any dimension $\nu \geq 2$). This is identical with the rate of decay obtained by time-independent methods.

It is easily seen from (6.28) that the l.h.s. of (6.55) corresponds to the averaged total cross section for a beam having energy distribution $\lambda^{(\nu-1)/2}|\psi(\lambda)|^2$ (in other words the vector g in (6.28) has the form $(\mathcal{U}_o g)_\lambda(\underline{\omega}) = \lambda^{(\nu-1)/4}\psi(\lambda)h(\underline{\omega})$). The bound (6.55) then has the interesting feature that it is the product of a term depending only on the interaction and a term depending only on properties of the beam.

So far we have assumed the potential to be square-integrable. By choosing in (6.37) an operator $J \neq I$, one may also treat other classes of potentials. We consider here the case where v may have arbitrary local singularities in a bounded subset of $\mathbb{R}^\nu$. More precisely, let us consider an infinitely differentiable function $j : \mathbb{R}^\nu \to \mathbb{R}$ such that $j(\underline{x}) = 0$ on some ball B_r and $j(\underline{x}) = 1$ in the complement of some larger ball B_ρ. If J denotes the multiplication operator by $j(\underline{x})$, then (6.37) becomes, by using (2.33) :

$$||RP(\psi)||_{HS} \leq \int_{-\infty}^{\infty} ||VJU_t^oP(\psi)||_{HS}dt$$

$$+ \int_{-\infty}^{\infty} ||[(\Delta J)(\underline{Q}) + 2i\sum_{m=1}^{\nu} j_{,m}(\underline{Q})P_m]U_t^oP(\psi)||_{HS}dt. \qquad (6.60)$$

If $(1+|\underline{x}|)^{1/2}[\log(2+|\underline{x}|)]^{(1+\varepsilon)/2}v(\underline{x})j(\underline{x})\in L^2(\mathbb{R}^\nu)$, with $\varepsilon > 0$, the first term on the r.h.s. is finite and may be estimated as before. Its contribution to the bound for $\int\lambda^{(\nu-1)/2}|\psi(\lambda)|^2\bar{\sigma}(\lambda)d\lambda$ is given by the r.h.s. of (6.55), with v replaced by vj. However there will be another contribution to this bound, arising from the second term on the r.h.s. of (6.60). This contribution does not depend explicitly on the potential v. It depends only on the cut-off function j (hence indirectly on v, in the sense that j must vanish at the singularities of v). We shall estimate this contribution in Lemma 6.16 below; in particular we shall show that it is always finite. Anticipating this result, we see that the finiteness of $\bar{\sigma}(\lambda)$ a.e. (as well as the existence of the wave operators) depends only on the behaviour of the potential near infinity and is completely independent of the details of its local behaviour. In particular the result of Corollary 6.15 is true without the hypothesis that $v\in L^2_{loc}(\mathbb{R}^\nu)$.

If v is of finite range, i.e. $v(\underline{x}) = 0$ for all $|\underline{x}| \geq r_o$, then the first term on the r.h.s. of (6.60) is zero if $j(\underline{x}) = 0$ on B_{r_o}. In this case one obtains a bound for $\int\lambda^{(\nu-1)/2}|\psi(\lambda)|^2\bar{\sigma}(\lambda)d\lambda$ which depends only on the range r_o of v but which is completely independent of the behaviour of v on B_{r_o} (in particular of the coupling constant). We shall show that for large r_o this bound has the same form as the (exact) expression for the total scattering cross section in classical mechanics (e.g. for a spherically symmetric potential of range r_o), i.e. it is proportional to $r_o^{\nu-1}$.

<u>Lemma 6.16</u> : Let $j : \mathbb{R}^\nu \to \mathbb{R}$ be bounded and continuously differentiable, and let ψ be as in Lemma 6.11. Then, for any $\kappa,\delta\in\mathbb{R}$ (in the notation introduced after Equation (2.32)) :

$$I_o := [\int_{-\infty}^{\infty} \|(\Delta j)(\underline{Q})U_t^o P(\psi)\|_{HS} dt]^2 \le \frac{1}{16}(2\pi)^{2-\nu}\Theta_\nu \|\psi_{\nu-2}\| \cdot$$

$$\cdot [4\|\psi_{\nu-4}\| \, \|\Delta j\|^{2-\kappa} \| \, |\underline{Q}|(\Delta j)\|^{\delta} + \nu^{-1}\|\psi_{\nu-4}\| \, \|\Delta j\|^{\kappa} \| \, |\underline{Q}|(\Delta j)\|^{2-\delta}$$

$$+ 4\|\psi'_{\nu-2}\|^2 \|\psi_{\nu-4}\|^{-1} \|\Delta j\|^{2+\kappa} \| \, |\underline{Q}|(\Delta j)\|^{-\delta}], \qquad (6.61)$$

$$[\int_{-\infty}^{\infty} \|j_{,m}(\underline{Q})P_m U_t^o P(\psi)\|_{HS} dt]^2 \le [16\nu(\nu+2)]^{-1}(2\pi)^{2-\nu}\Theta_\nu \|\psi_\nu\| \, \|j_{,m}\| \cdot$$

$$\cdot [(4\nu+11)\|\psi_{\nu-2}\| \, \| \, |\underline{Q}| j_{,m}\| +$$

$$+ (4\nu+8)\|\psi'_\nu\|^2 \|\psi_{\nu-2}\|^{-1} \|j_{,m}\|^2 \| \, |\underline{Q}| j_{,m}\|^{-1}]. \qquad (6.62)$$

Proof : We apply the Schwarz inequality as in the proof of Proposition 6.14 to get that for any $\gamma > 0$

$$I_o \le \int_{-\infty}^{\infty} (1+\gamma^2 s^2)^{-1} ds \cdot \int_{-\infty}^{\infty} (1+\gamma^2 t^2) \|(\Delta j)(\underline{Q})U_t^o P(\psi)\|_{HS}^2 dt.$$

The integral over ds is equal to $\pi\gamma^{-1}$. The integral over dt may be calculated from (6.39) and (6.41), by taking $w(\underline{x}) = (\Delta j)(\underline{x})$ in (6.39) and $w(\underline{x}) = (1+|\underline{x}|)(\Delta j)(\underline{x})$ in (6.41). To arrive at the r.h.s. of (6.61), we have set $\gamma = \|\psi_{\nu-2}\| \, \|\psi_{\nu-4}\|^{-1} \|\Delta j\|^{\kappa} \| \, |\underline{Q}|(\Delta j)\|^{-\delta}$.

Similarly one obtains (6.62) from (6.40) and (6.42). To get (6.62) we have majorized the last bracket in (6.42) by $3\| \, |\underline{Q}|(I+|\underline{Q}|)^{-1} w\|^2$ and taken $\gamma = \|\psi_\nu\| \, \|\psi_{\nu-2}\|^{-1} \|j_{,m}\| \, \| \, |\underline{Q}| j_{,m}\|^{-1}$.□

If $j(\underline{x}) = 1$ near infinity, then $j_{,m}$ and Δj are zero near infinity, so that $j_{,m}, \Delta j, |\underline{Q}| j_{,m}$ and $|\underline{Q}|(\Delta j)$ all belong to $L^2(\mathbb{R}^\nu)$. This shows that the bounds (6.61) and (6.62) are finite in this case, as stated before.

In the same way as for non-singular potentials, we may use Lemma 6.16 to estimate the high energy behaviour of the total cross section for singular potentials. If we take ψ of the form indicated after Corollary 6.15, we obtain instead of (6.59) a bound of the form

$$\int_{(k_0+b)^2}^{(k_0+c)^2} \lambda^{(\nu-2)/2}\bar{\sigma}(\lambda)d\lambda \leq \kappa_\nu(\varphi,vj,j)k_0^{\nu-1}, \qquad (6.63)$$

where κ_ν is a constant depending on φ, vj and j. As in (6.59), the contribution from (6.61) to the r.h.s. of (6.63) behaves like $k_0^{\nu-3}$, whereas that from (6.62) behaves like $k_0^{\nu-1}$ and therefore dominates (this is due to the additional factor P_m on the l.h.s. of (6.62)). This implies essentially that $\bar{\sigma}(\lambda) = 0(1)$ as $\lambda \to \infty$ for potentials having strong local singularities.

Finally we use Lemma 6.16 to obtain a bound on the scattering cross section depending only on the range r_0 of the potential v.

Proposition 6.17 : Let $v : \mathbb{R}^\nu \to \mathbb{R}$ be such that $v(\underline{x}) = 0$ for $|\underline{x}| \geq r_0$, let H be an arbitrary self-adjoint extension of the operator $\hat{H} = -\Delta + v(\underline{x})$ defined on $D(\hat{H}) = C_0^\infty(\mathbb{R}^\nu\backslash B_{r_0})$, and let ψ be as in Lemma 6.11. Then there is a number $c_\nu(\psi)$ such that for all $r_0 > 0$:

$$\int_0^\infty \lambda^{(\nu-1)/2}|\psi(\lambda)|^2\bar{\sigma}(\lambda)d\lambda \leq c_\nu(\psi)[r_0^{\nu-1} + \min(1,r_0^{\nu-4})]. \qquad (6.64)$$

Proof : Let $f : \mathbb{R}^\nu \to \mathbb{R}$ be infinitely differentiable and such that $f(\underline{x}) = 0$ for $|\underline{x}| \leq 1$ and $f(\underline{x}) = 1$ for $|\underline{x}| \geq 2$. Set $j(\underline{x}) = f(r_0^{-1}\underline{x})$. We then have

$$(\Delta j)(\underline{x}) = r_0^{-2}(\Delta f)(r_0^{-1}\underline{x}), \quad j_{,m}(\underline{x}) = r_0^{-1}(f_{,m})(r_0^{-1}\underline{x}).$$

Consequently

$$\begin{aligned} ||\Delta j||^2 &= r_0^{-4} \int |(\Delta f)(r_0^{-1}\underline{x})|^2 d^\nu x = r_0^{-4+\nu} \int |(\Delta f)(\underline{y})|^2 d^\nu y \\ &= ||\Delta f||^2 r_0^{\nu-4}. \qquad (6.65) \end{aligned}$$

Similarly one gets that

$$\| \, |\underline{Q}| (\Delta j) \|^2 = \| \, |\underline{Q}| (\Delta f) \|^2 r_0^{\nu-2}, \tag{6.66}$$

$$\| j_{,m} \|^2 = \| f_{,m} \|^2 r_0^{\nu-2}, \tag{6.67}$$

$$\| \, |\underline{Q}| j_{,m} \|^2 = \| \, |\underline{Q}| f_{,m} \|^2 r_0^{\nu}. \tag{6.68}$$

Setting $\kappa = \nu - 2$ and $\delta = \nu - 4$ in (6.61), we obtain from (6.65) and (6.66) that the r.h.s. of (6.61) has the form

$$c'_\nu(\psi)[r_0^{\nu-4} + r_0^{\nu-2} + r_0^{\nu-4}].$$

On the other hand, by using (6.67) and (6.68), we see that the r.h.s. of (6.62) has the form

$$c''_\nu(\psi)[r_0^{\nu-1} + r_0^{\nu-3}].$$

In view of (6.36) and (6.60), this implies that for all $r_0 > 0$

$$\int_0^\infty \lambda^{(\nu-1)/2} |\psi(\lambda)|^2 \bar{\sigma}(\lambda) d\lambda \leq c_\nu(\psi)[r_0^{\nu-1} + r_0^{\nu-4}]. \tag{6.69}$$

The term in $r_0^{\nu-4}$ dominates for $r_0 < 1$. For these values of r_0 we may take a different cut-off function j which is independent of r_0, for instance $j(\underline{x}) = f(\underline{x})$. The r.h.s. of (6.61) and (6.62) is then independent of r_0, i.e. the integral in (6.69) may be bounded by a constant independent of r_0 for $0 < r_0 < 1$. This implies together with (6.69) the bound (6.64). □

Notice that for large r_0 the bound (6.64) has the form $cr_0^{\nu-1}$. For $r_o \to 0$, the bound tends to zero like $r_0^{\nu-4}$ if $\nu \geq 5$, whereas it remains constant if $\nu \leq 4$.

APPENDIX

We collect here a few definitions and results from the theory of L^p-spaces. Let $(M;\mu)$ be a measure space, in other words let μ be a measure defined on a σ-algebra R of subsets of the set M [R, p. 217]. If Δ is a measurable subset of M (i.e. an element of R), we denote by χ_Δ the <u>characteristic function</u> of Δ, which is defined as follows :

$$\chi_\Delta(s) = \begin{cases} 1 & \text{if } s\in\Delta \\ 0 & \text{if } s\notin\Delta. \end{cases} \tag{A.1}$$

For $p\in[1,\infty]$, $L^p(M;d\mu)$ is the set of all equivalence classes of measurable function $f : M \to \mathbb{C}$ satisfying $\|f\|_p < \infty$, where two functions are said to be equivalent if they are equal μ-almost everywhere, and where $\|f\|_p$ is defined as follows :

$$\|f\|_p := [\int_M |f(s)|^p d\mu(s)]^{1/p} \quad \text{if } p < \infty \tag{A.2}$$

and

$$\|f\|_\infty := \operatorname*{ess\,sup}_{s\in M} |f(s)|. \tag{A.3}$$

Here ess sup $g(s)$ is the infimum of sup $h(s)$ as h varies over all functions that are equal to g almost everywhere. In other words ess sup $g(s)$ is the infimum of all m such that the measure $\mu(\Delta_m)$ of the set $\Delta_m = \{s\in M | g(s) > m\}$ is zero. If for example M is the closed interval $M = [a,b]$, μ Lebesgue measure and f is continuous on $[a,b]$, then $\|f\|_\infty = \max_{x\in[a,b]} |f(x)|$.

$L^p(M;d\mu)$ is a complete normed linear space with respect to the norm $\|\cdot\|_p$. If $f\in L^p(M;d\mu)$, $g\in L^q(M;d\mu)$ and $1/r = 1/p + 1/q$, then $f(\cdot)g(\cdot)\in L^r(M;d\mu)$ and

$$\|fg\|_r \le \|f\|_p \|g\|_q. \tag{A.4}$$

This is known as the <u>Hölder inequality</u>. Its proof is simple and may be found e.g. in [R, p. 113]. The following facts about L^p-spaces are often useful :

<u>Lemma A.1</u> : (a) If Δ is a measurable subset of M with finite measure, $p\in[1,\infty]$ and $f\in L^p(M;d\mu)$, then $\chi_\Delta(\cdot)f(\cdot)\in L^r(M;d\mu)$ for each $r\in[1,p]$.

(b) If $1 \le p < q \le \infty$, then $L^p(M;d\mu) \cap L^q(M;d\mu) \subseteq L^r(M;d\mu)$ for each $r\in[p,q]$.

<u>Proof</u> : (a) This follows from the Hölder inequality. If $q = (r^{-1} - p^{-1})^{-1}$, then $q\in[r,\infty]$ and $\|\chi_\Delta f\|_r \le \|\chi_\Delta\|_q \|f\|_p$, which is finite since $\|\chi_\Delta\|_q = [\mu(\Delta)]^{1/q} < \infty$ ($\mu(\Delta) \equiv$ measure of Δ).

(b) If $f\in L^p \cap L^q$, write $f = f_1 + f_2$ with

$$f_1(s) = \begin{cases} f(s) & \text{if } |f(s)| \le 1 \\ 0 & \text{if } |f(s)| > 1. \end{cases}$$

For $r \ge p$, we have $|f_1(s)|^r \le |f_1(s)|^p$, hence $f_1\in L^r$ for each $r \ge p$. Furthermore, let $\Delta = \{s\in M \mid |f(s)| > 1\}$. Then $\mu(\Delta) < \infty$, since $f\in L^p$ with $p < \infty$. Consequently $\chi_\Delta f = f_2 \in L^r$ for each $r \le q$, by (a). Hence $f_1 + f_2 \in L^r$ for each $r\in[p,q]$. □

An important theorem, which allows one to interchange a limit with an integral, is the Lebesgue Dominated Convergence Theorem. We use it only for $p = 1$ [R] :

Lebesgue Dominated Convergence Theorem : Assume

i) $g, f_t \in L^1(M; d\mu)$ $(t \in \mathbb{R})$,

ii) $|f_t(s)| \leq g(s)$ for almost all $s \in M$ and all t,

iii) $\lim_{t \to t_0} f_t(s) = f(s)$ for almost all $s \in M$.

Then $f \in L^1(M; d\mu)$ and

$$\lim_{t \to t_0} \int_M f_t(s) d\mu(s) = \int_M f_{t_0}(s) d\mu(s).$$

Instead of using functions with values in $\mathbb{C}$, one could also consider functions from M to some normed space H_0 and define the spaces $L^p(M, H_0; d\mu)$. We use this in Chapters 5 and 6 for the case $p = 2$, with H_0 a Hilbert space. The relevant definitions are given in Section 5.4.

We now consider some special classes of functions in the space $L^p(\mathbb{R}^\nu)$, where μ is Lebesgue measure and $1 \leq p < \infty$. $S(\mathbb{R}^\nu)$ is the set of all infinitely differentiable functions $f : \mathbb{R}^\nu \to \mathbb{C}$ such that f and all its partial derivatives of any order tend to zero faster than any negative power of $|\underline{x}|$ as $|\underline{x}| \to \infty$. In symbols, f is in $S(\mathbb{R}^\nu)$ if, for each 2ν-tupel $i_1, \ldots, i_\nu$, $m_1, \ldots, m_\nu$ of non-negative integers, one has

$$\sup_{\underline{x} \in \mathbb{R}^\nu} \left| (x_1)^{i_1} \cdots (x_\nu)^{i_\nu} \frac{\partial^{m_1 + \cdots + m_\nu}}{\partial x_1^{m_1} \cdots \partial x_\nu^{m_\nu}} f(x_1, \ldots, x_\nu) \right| < \infty. \quad \text{(A.5)}$$

Lemma A.2 : (a) $S(\mathbb{R}^\nu)$ is dense in $L^p(\mathbb{R}^\nu)$, $1 \leq p < \infty$.

(b) $S(\mathbb{R}^\nu)$ is invariant under the Fourier transformation, i.e. $f \in S(\mathbb{R}^\nu) \Rightarrow \tilde{f} \in S(\mathbb{R}^\nu)$.

Proof : (a) follows from Lemma A.3 below. For (b), set

$$A_{i_1\cdots i_\nu m_1\cdots m_\nu}(\underline{k}) := k_1^{i_1}\cdots k_\nu^{i_\nu}\frac{\partial^{m_1+\cdots+m_\nu}}{\partial k_1^{m_1}\cdots\partial k_\nu^{m_\nu}}\tilde{f}(\underline{k})$$

and notice that this is the Fourier transform of

$$i^{i_1+\cdots+i_\nu+m_1+\cdots+m_\nu}\frac{\partial^{i_1+\cdots+i_\nu}}{\partial x_1^{i_1}\cdots\partial x_\nu^{i_\nu}}x_1^{m_1}\cdots x_\nu^{m_\nu}f(\underline{x}),$$

the absolute value of which is bounded, as a consequence of (A.5), by $c_m(1+|\underline{x}|)^{-m}$ for any $m > 0$ [AJS, p. 33]. Hence, taking $m > \nu$:

$$|A_{i_1\cdots i_\nu m_1\cdots m_\nu}(\underline{k})| \le (2\pi)^{-\nu/2}\, c_m\int(1+|\underline{x}|)^{-m}d^\nu x < \infty. \quad \square$$

Next we introduce $C_0^\infty(\mathbb{R}^\nu)$ as the set of all infinitely differentiable functions f from $\mathbb{R}^\nu$ to $\mathbb{C}$ each of which is identically zero outside some bounded subset of $\mathbb{R}^\nu$ (depending on f). Clearly $C_0^\infty(\mathbb{R}^\nu)$ is a subset of $S(\mathbb{R}^\nu)$, so that the Fourier transform of a function in $C_0^\infty(\mathbb{R}^\nu)$ is infinitely differentiable and rapidly decreasing at infinity.

Finally, if Γ is a closed subset of $\mathbb{R}^\nu$ of measure zero, we denote by $C_0^\infty(\mathbb{R}^\nu\backslash\Gamma)$ the set of all functions in $C_0^\infty(\mathbb{R}^\nu)$ with supports contained in $\mathbb{R}^\nu\backslash\Gamma$. We recall that the support supp f of a function f is defined as the closure of the set $\{\underline{x}\in\mathbb{R}^\nu | f(\underline{x}) \neq 0\}$.

Lemma A.3 : Let $\Gamma\subset\mathbb{R}^\nu$ be closed and of Lebesgue measure zero. Then $C_0^\infty(\mathbb{R}^\nu\backslash\Gamma)$ is dense in $L^p(\mathbb{R}^\nu)$ for each $p\in[1,\infty)$. In particular $C_0^\infty(\mathbb{R}^\nu)$ is dense in $L^p(\mathbb{R}^\nu)$, $1 \le p < \infty$.

Idea of proof : Let $\varepsilon > 0$ and $f\in L^p(\mathbb{R}^\nu)$. Let Δ be an open set such that $\Gamma\subset\Delta$ and $||\chi_\Delta f||_p < \varepsilon/3$. Let $R < \infty$ be such that $\int_{|\underline{x}|>R}|f(\underline{x})|^p d^\nu x < (\varepsilon/3)^p$. Define f_0 by

$$f_o(\underline{x}) = \begin{cases} f(\underline{x}) & \text{if } \underline{x} \notin \Delta \text{ and } |\underline{x}| \le R \\ 0 & \text{otherwise.} \end{cases}$$

Then $\|f - f_o\|_p < 2\varepsilon/3$, and f_o has compact support in $\mathbb{R}^\nu \backslash \Gamma$.

Choose a function $\varphi \in C_o^\infty(\mathbb{R}^\nu)$ such that $\varphi \ge 0$, supp $\varphi \subseteq B_1 \equiv \{\underline{x} \in \mathbb{R}^\nu \,|\, |\underline{x}| \le 1\}$ and $\int \varphi(\underline{x}) d^\nu x = 1$. For $\eta > 0$, define $f_{o,\eta}$ by

$$f_{o,\eta}(\underline{x}) = \int \varphi(\underline{y}) f_o(\underline{x} - \eta \underline{y}) d^\nu y = \eta^{-\nu} \int \varphi(\eta^{-1}(\underline{x} - \underline{z})) f_o(\underline{z}) d^\nu z. \quad \text{(A.6)}$$

Then $f_{o,\eta}$ is infinitely differentiable. Furthermore, if η is smaller than the distance from supp f_o to Γ, the support of $f_{o,\eta}$ is contained in $(\mathbb{R}^\nu \backslash \Gamma) \cap B_{R+\eta}$. Hence $f_{o,\eta} \in C_o^\infty(\mathbb{R}^\nu \backslash \Gamma)$. A short calculation shows that $\|f_o - f_{o,\eta}\|_p \to 0$ as $\eta \to 0$, which proves the lemma. (For more details, we refer to [S, §I.2.4] or [H, p. 3].) □

<u>Lemma A.4</u> : Let $\Gamma \subset \mathbb{R}^\nu$ be closed and of measure zero.

(a) Let Δ be a compact subset of $\mathbb{R}^\nu \backslash \Gamma$. Then there is a function $\varphi \in C_o^\infty(\mathbb{R}^\nu \backslash \Gamma)$ such that $0 \le \varphi \le 1$ and $\varphi(\underline{x}) = 1$ for all $\underline{x}$ in some neighbourhood of Δ.

(b) If $f \in C_o^\infty(\mathbb{R}^\nu \backslash \Gamma)$, there is a function $\varphi \in C_o^\infty(\mathbb{R}^\nu \backslash \Gamma)$ such that $\varphi f = f$.

<u>Proof</u> : Let δ be the distance from Δ to Γ. Let Δ' be the set of all points at distance $\le \frac{1}{2}\delta$ from Δ. Define $\chi_{\Delta',\eta}$ as in (A.6) by replacing f_o by the characteristic function $\chi_{\Delta'}$ of Δ'. Then, for $\eta < \delta/4$, $\varphi = \chi_{\Delta',\eta}$ has all required properties. (b) follows from (a) by taking Δ = supp f. □

We end with a result on uniformly continuous functions. A function $\varphi : \mathbb{R} \to \mathbb{C}$ is said to be <u>uniformly continuous</u> if, given any $\varepsilon > 0$, there is a $\delta > 0$ such that $|\varphi(t+\tau) - \varphi(t)| < \varepsilon$ for all $t \in \mathbb{R}$ and all $\tau \in (-\delta, \delta)$.

Lemma A.5 : Suppose $\varphi : [a,\infty) \to \mathbb{C}$ is uniformly continuous and $\int_a^\infty |\varphi(t)|^p dt < \infty$ for some $p \in [1,\infty)$. Then $\lim \varphi(t) = 0$ as $t \to +\infty$. A similar result holds near $-\infty$.

Proof : Assume that $\varphi(t) \not\to 0$ as $t \to +\infty$. Then there is an $\varepsilon > 0$ and a sequence $\{t_n\}$ such that $t_n \to +\infty$, $t_{n+1} - t_n \geq 2$ and $|\varphi(t_n)| \geq \varepsilon$. By the uniform continuity of φ, there is a $\eta \in (0,1)$ such that $|\varphi(t)| \geq \varepsilon/2$ for all $t \in (t_n - \eta, t_n + \eta)$ and all n. Hence

$$\int |\varphi(t)|^p dt \geq \sum_{n=1}^{\infty} \int_{t_n-\eta}^{t_n+\eta} |\varphi(t)|^p dt \geq \sum_{n=1}^{\infty} 2\eta(\varepsilon/2)^p = \infty,$$

which contradicts the second hypothesis of the lemma. Hence $\varphi(t) \to 0$ as $t \to +\infty$. □

NOTES

Chapter 1

A. In Chapter 1 we give only those notions and results from Hilbert space theory that are needed at the later stages of the text. For a more comprehensive and simple introduction to the theory of Hilbert spaces and linear operators we recommend the book by Akhiezer and Glazman [AG]. More advanced texts are [K], [RN], [RS]. For details about measure theory and the L^p-spaces we refer to [R] and [RN].

B. With regard to Remark 1.25, we wish to point out that various more general types of Banach-space valued integrals have been studied. See [HP] for details.

Chapter 2

A. If A is not semi-bounded, the criterion for essential self-adjointness given in Proposition 2.4 does not apply. One can then use the following criterion (see e.g. [AJS, §2-3]) :

Proposition N.1 : The symmetric operator A is essentially self-adjoint if and only if $\mathcal{R}(A + iI)$ and $\mathcal{R}(A - iI)$ are dense in $\mathcal{H}$.

This can for instance be used to show that P_m and Q_m ($m = 1,\ldots,\nu$) are essentially self-adjoint on $\mathcal{S}(\mathbb{R}^\nu)$, by using an argument similar to that in the proof of Proposition 2.18.

The following is an essential self-adjointness criterion for the restriction of the infinitesimal generator of an evolution group to certain subsets of its domain (see e.g. [RS, Theorem VIII.11]) :

Proposition N.2 : Let $\{U_t\}$ be an evolution group and A its infinitesimal generator. If $\mathcal{D}$ is a dense linear subset of D(A) such that $U_t\mathcal{D}\subseteq\mathcal{D}$ for all $t\in\mathbb{R}$, then the restriction of A to $\mathcal{D}$ is essentially self-adjoint.

B. By Example 2.26, it is crucial to require in Proposition 2.23 that B be A-bounded with A-bound υ *strictly less than 1* in order to deduce the self-adjointness of A + B. If B is A-bounded with A-bound $\upsilon = 1$, it is however still possible to prove that A + B is essentially self-adjoint on D(A) (see [K, p. 289 and 571]).

C. In Section 2.5 we required that the potential be locally L^2 away from a closed set of measure zero, which ensures that the minimal operator is densely defined, hence symmetric. This local L^2-condition can in some cases be weakened to a local L^1-condition by working with quadratic forms. The operator $H_0 = \underline{P}^2$ determines a quadratic form q_0 by

$$q_0(f,g) = (|\underline{P}|f,|\underline{P}|g),$$

where $f,g\in D(|\underline{P}|)$ and $(\mathcal{F}|\underline{P}|f)(\underline{k}) := |\underline{k}|\tilde{f}(\underline{k})$. Similarly, a potential v determines a quadratic form q_v by

$$q_v(f,g) = (|V|^{1/2}f,V^{1/2}g),$$

where $f,g\in D(|V|^{1/2})$ and $|V|^{1/2}$ and $V^{1/2}$ are the multiplication operators by $|v(\underline{x})|^{1/2}$ and $|v(\underline{x})|^{1/2}\cdot\text{sign}\, v(\underline{x})$ respectively. Notice that $v\in L^1_{loc} \Rightarrow |v|^{1/2}\in L^2_{loc}$.

Instead of adding the operators H_0 and V, one may add the quadratic forms q_0 and q_v. Under a smallness assumption on the negative part of v, one can show that the sum of the two forms is the quadratic form corresponding to a uniquely determined self-adjoint operator H, which is then taken as the Hamiltonian. For details we refer to [K], [RS]. Applications to scattering theory may be found e.g. in [RS], [4], [19] and [20].

D. Domain properties of Schrödinger operators with singular potentials were first studied in detail by Ikebe and Kato [21]. They use a method involving Green's functions which leads to a slightly stronger form of the results given in Section 2.5 Another proof of these results which is more closely related to that given here, but which uses also complex interpolation, may be found in [SH, Lemma 9.2.2]. See also [11], [20], [22].

Chapter 4

A. Strongly continuous unitary one-parameter groups in a Hilbert space are a special case of semi-groups in Banach spaces, about which there is a considerable literature (see [HP], [K], [RS]). For further developments on ergodic theory we refer to [HO], [J].

B. The result of Lemma 4.17 is related to Weyl's criterion characterizing the essential spectrum of a self-adjoint operator A. If A has purely continuous spectrum, then Weyl's criterion states that $\mu \in \sigma(A) \iff \exists \{g_n\} \in D(A)$ such that $\|g_n\| = 1$, w-lim $g_n = \theta$ and s-lim$(A-\mu)g_n = \theta$ as $n \to \infty$. For a general A, its essential spectrum is defined as $\sigma(A) \backslash \sigma_d(A)$, where the discrete spectrum $\sigma_d(A)$ consists of all eigenvalues λ of A that are of finite multiplicity and such that the distance from λ to $\sigma(A)\backslash\{\lambda\}$ is strictly positive. Weyl's criterion then says that μ belongs to the essential spectrum of A if and only if there is a sequence $\{g_n\}$ having the above-mentioned properties [AJS, Lemma 5.19].

C. The existence, for each $f \in H$, of a function $\mu_f : \mathbb{R} \to [0,\infty)$ having the properties stated in Remark 4.31 can be used to prove the spectral theorem for self-adjoint operators. This theorem states that, given a self-adjoint operator A, or equivalently an evolution group $\{U_t\}$, there is a uniquely determined family of projections $\{E_\lambda\}_{\lambda\in\mathbb{R}}$ satisfying :

i) $E_\lambda E_\mu = E_\mu E_\lambda = E_{\min\{\lambda,\mu\}} \quad \forall \lambda,\mu\in\mathbb{R}$,

ii) $\underset{\varepsilon\to+0}{\text{s-lim}}\, E_{\lambda+\varepsilon} = E_\lambda \quad \forall \lambda\in\mathbb{R}$,

iii) $\underset{\lambda\to-\infty}{\text{s-lim}}\, E_\lambda = 0$ and $\underset{\lambda\to+\infty}{\text{s-lim}}\, E_\lambda = I$,

and such that

$$(f,U_t g) = \int_{-\infty}^{\infty} e^{-i\lambda t} d(f,E_\lambda g) \qquad \forall f,g\in H.$$

The function μ_f is related to the family of projections $\{E_\lambda\}$ by

$$\mu_f(\lambda) = (f,E_\lambda f).$$

For a simple approach to the spectral theorem from this point of view, see [HO].

Chapter 5

A. The definition of bound states and scattering states in terms of the position probability density was introduced by Ruelle [23], who proved for a class of two-body and many-body systems that $M_0(H) = H_p(H)$ and $\overline{M}_\infty^{\pm}(H) = H_c(H)$. Generalizations and further developments were obtained in [19] and [2]. In [2] it is shown that, if $\int_{-\infty}^{\infty} \|F_r U_t f\|^2 dt < \infty$ for each $r\in(0,\infty)$, (in other words, if the "time of sojourn" in each ball B_r is finite), then $f\in H_{ac}(H)$.

An abstract scattering theory, starting from a definition of the wave operators on the subspaces of scattering states $M_\infty^{\pm}(H_0)$ rather than, as was done in earlier work, on $H_{ac}(H_0)$, has been developed in [24]. Wave operators defined on the subspace of continuity $H_c(H_0)$ are studied in [25].

B. The condition for the existence of the wave operators given in Proposition 5.19 is often referred to as the Cook criterion [26]. In recent years various generalizations of this criterion have been obtained. That given in Proposition 5.20 is due to Kato [27]. Further references may be found in [27].

C. Various methods have been developed for proving asymptotic completeness. One of these uses properties of trace class operators; most of the other methods somehow involve a control over the behaviour of the resolvent $(H - \lambda \pm i\varepsilon)^{-1}$ of the total Hamiltonian H near its spectrum, often referred to as the "limiting absorption principle" (λ is real, ε is positive but arbitrarily small). We refer to the two recent textbooks [AJS] and [RS, Vols III,IV] for details and references.

The method given in this text, which is essentially time-dependent and closer to intuition than previous methods, was discovered by Enss [28] and announced in 1978. Various improvements and generalizations have been obtained since then, see [29]-[33], [12]-[14] and [10]. The presentation of the abstract part in Section 5.3 is essentially based on [10], the application to Schrödinger operators in Section 5.4 is closely related to the method of [12] and [14]. More general free Hamiltonians are considered in particular in [31] and [33].

D. Results on generalized asymptotic completeness have been obtained by different methods in [11], [20] and [5].

E. As pointed out in Remark 5.36 (b), one has to select suitable operators $J_\pm$ different from the identity operator I and not satisfying (C8) in order to obtain wave operators $W_\pm =$ s-lim $U_t^* J_\pm U_t^o$ for Schrödinger operators with long range potentials. Instead of replacing I by $J_\pm$ in the definition of the usual wave operators $\Omega_\pm$, one may also replace the group $\{U_t^o\}$ by two more complicated families of unitary operators $\{T_t^{\pm}\}$, which may be constructed from the potential v. In this case the wave operators are given by $W_{\pm} = \text{s-lim}\ U_t^* T_t^{\pm}$ as $t \to \pm\infty$. Details on this latter approach may be found for instance in [15], [34], [35], [30], [AJS, Chapter 13] and [RS, §XI.9].

A convenient choice of operators $J_{\pm}$ different from I but satisfying (C8) also allows one to prove the existence and completeness of the usual wave operators $\Omega_{\pm} = \text{s-lim}\ U_t^* U_t^o$ for potentials that are not of short range but sufficiently rapidly oscillating near infinity (for example, in $L^2(\mathbb{R}^3)$: $v(r) = cr^{\alpha}\sin(r^{\beta})$ with $\alpha \in \mathbb{R}$ and $\beta > \alpha + 2$, where $r = |\underline{x}|$). See [10], [36], [37] and the references given in [36].

Chapter 6

A. The method described in Section 6.3 for deriving bounds on the averaged total scattering cross section was first introduced in [37]. The estimates given here slightly improve the results of [37]. Similar bounds can be obtained for potentials that are rapidly oscillating near infinity, even with diverging amplitude, such as $v(\underline{x}) = c|\underline{x}|^{\alpha}\sin(|\underline{x}|^{\beta})$ in $L^2(\mathbb{R}^3)$, with $\alpha \in \mathbb{R}$ and $\beta > \alpha + 3$ [37].

As a function of the coupling constant g, our bounds are independent of g for potentials of compact support and proportional to g^2 for large g in the general case. On the other hand one can derive a bound like g^{α} for large g, where α depends on the rate of decay of the potential at infinity [38].

Corollary 6.15, giving a sufficient condition on the rate of decay of the potential for the finiteness of the total scattering cross section, is not completely optimal. In [40] it has been shown by time-independent methods that $\bar{\sigma}(\lambda) < \infty$ if $|v(\underline{x})| \leq$

$\leq c(1+|\underline{x}|)^{-(\nu+1)/2}[\log(2+|\underline{x}|)]^{-\frac{1}{2}-\varepsilon}$ for some $\varepsilon > 0$, and for $\nu = 3$. If $\varepsilon = 0$, one may have $\bar{\sigma}(\lambda) = \infty$ for all λ [38].

As in Section 5.3 (cf. Propositions 5.19 and 5.20) it is possible to use the operator $Y_z = JR_z^o - R_z J$ instead of $HJ - JH_o$ in the bounds for the scattering cross section. Using the fact that $(H_o - z)^{-2}R = -iW_+^* \int U_t^* Y_z U_t^o dt$, one easily finds that

$$\int_0^\infty (\lambda - z)^{-2}|\psi(\lambda)|^2 \|R(\lambda)\|_{HS}^2 = \|(H_o - z)^{-2}RP(\psi)\|_{HS}^2$$

$$\leq [\int_{-\infty}^\infty \|Y_z U_t^o P(\psi)\|_{HS} dt]^2.$$

Finally we mention that a different time-dependent method for proving the finiteness of $\bar{\sigma}(\lambda)$ was already used in [39]. This approach is based on trace class properties of $R(\lambda)$ but does not easily give simple bounds on the cross section.

B. In [38] a somewhat different time-dependent method has been developed for obtaining bounds on the total scattering cross section for fixed initial direction $\underline{\omega}_o$, but averaged over a range of energies. The idea of this method is as follows. Choose the x_ν-axis in the direction of $\underline{\omega}_o$, and consider a state of the form $g_r = f(x_\nu)h_r(x_1,\ldots,x_{\nu-1})$, where $\tilde{f}$ has support in a small interval of the positive k-axis (it represents the momentum distribution of the incoming state), whereas $\{h_r\}$ is a sequence of functions tending to 1 as $r \to \infty$, i.e. $\tilde{h}_r$ tends to a multiple of the $(\nu-1)$-dimensional δ-function. Let C be a cone with apex at the origin and χ_C its characteristic function. Then, by a calculation as in (6.29) :

$$\|\chi_C(\underline{P})Rg_r\|^2 = \int d\lambda \int_{C_u} d\omega \Big|\int d\omega' r(\lambda;\underline{\omega},\underline{\omega}')\frac{1}{\sqrt{2}}\lambda^{(\nu-2)/4}\tilde{g}_r(\sqrt{\lambda}\underline{\omega})\Big|^2.$$

Now, using the special form of $\tilde{g}_r$ and letting $r \to \infty$, we may, as in the proof of Proposition 6.8, set $d\omega' = k^{-(\nu-1)}dk_1\cdots dk_{\nu-1}$ and integrate away the δ-function. Thus (setting $k_r = k$ and $\lambda = k^2$)

$$\lim_{r\to\infty} \| \chi_C(\underline{P}) R g_r \|^2$$

$$= \int 2k\,dk \int_{C_u} d\omega \frac{1}{2} k^{\nu-2} \frac{(2\pi)^{\nu-1}}{k^{2(\nu-1)}} |r(k^2;\underline{\omega},\underline{\omega}_o)|^2 |\tilde{f}(k)|^2$$

$$= \int dk \int_{C_u} d\omega \frac{d\sigma}{d\omega}(k^2,\underline{\omega}_o;\underline{\omega}) |\tilde{f}(k)|^2 .$$

Taking $C = S^{\nu-1}\backslash\{\underline{\omega}_o\}$, this gives

$$\lim_{r\to\infty} \| R g_r \|^2 = \int dk\, \sigma_{tot}(k^2,\underline{\omega}_o) |\tilde{f}(k)|^2. \tag{N.1}$$

Writing R as an integral over time as in Section 6.3, we get

$$\int dk\, \sigma_{tot}(k^2,\underline{\omega}_o) |\tilde{f}(k)|^2 \le [\lim_{r\to\infty} \int_{-\infty}^{\infty} \| V U_t^o g_r \| dt]^2, \tag{N.2}$$

where we have taken $J = I$. Now $U_t^o g_r = (U_t^{o,1} f)(U_t^{o,\nu-1} g_r)$, where $U_t^{o,m}$ denotes the m-dimensional Schrödinger free evolution. For suitable functions g_r, the limit $r \to \infty$ can be carried out under the integral, and one is left with a *one-dimensional* estimate

$$\int dk\, \sigma_{tot}(k^2,\omega_o) |\tilde{f}(k)|^2 \le [\int_{-\infty}^{\infty} \| V_1 U_t^{o,1} f \| dt]^2, \tag{N.3}$$

where $V_1(x) = [\int dx_1 \cdots dx_{\nu-1} |V(x_1,\ldots,x_{\nu-1},x)|^2]^{1/2}$.

From (N.3) one obtains bounds on the total scattering cross section that are essentially equivalent to those of Section 6.3 but do not require an average over the initial direction [38].

C. An important theoretical quantity in quantum mechanics is the so-called scattering amplitude, the absolute square of which is just the differential scattering cross section. In the terminology of these notes, the scattering amplitude $f(\lambda;\underline{\omega}_o \to \underline{\omega})$ for scattering at energy λ from the initial direction $\underline{\omega}_o$ into the final direction $\underline{\omega}$ is given as follows

$$f(\lambda;\underline{\omega}_0 \rightarrow \underline{\omega}) = \left(\frac{-2\pi i}{\sqrt{\lambda}}\right)^{(\nu-1)/2} r(\lambda;\underline{\omega},\underline{\omega}_0). \tag{N.4}$$

D. A discussion of scattering into cones for scattering systems for which the free evolution group is not the Schrödinger free evolution group can be found in [41].

E. Finally we should mention that a fair number of similar results have also been derived for multiparticle quantum systems, for example self-adjointness of the Hamiltonian, existence of wave operators [AJS], [RS], scattering into cones [42], asymptotic completeness at low energies [29] and bounds on the total scattering cross section [38], [43]. What is still missing is a time-dependent proof of asymptotic completeness in the general case.

BIBLIOGRAPHY

[AG] AKHIEZER N.I. and GLAZMAN I.M. : Theory of Linear Operators in Hilbert Space (English Translation), Vols I and II, Frederick Ungar, New York (1963).

[AJS] AMREIN W.O., JAUCH J.M. and SINHA K.B. : Scattering Theory in Quantum Mechanics, Benjamin, Reading, Mass. (1977).

[B] BOCHNER S. : Lectures on Fourier Integrals, Princeton University Press (1959).

[GR] GUSTAFSON K. and REINHARDT W.P. : Classical, Semiclassical, and Quantum Mechanical Problems in Mathematics, Plenum, New York (1981).

[H] HÖRMANDER L. : Linear Partial Differential Operators, Springer, Berlin (1963).

[HO] HOPF E. : Ergodentheorie, Springer, Berlin (1970).

[HP] HILLE E. and PHILLIPS R.S. : Functional Analysis and Semi-Groups, AMS Colloq. Publ., Vol. 31, Providence, R.I. (1957).

[J] JACOBS K. : Neuere Methoden und Ergebnisse der Ergodentheorie, Springer, Berlin (1960).

[JA] JAUCH J.M. : Foundations of Quantum Mechanics, Addison-Wesley, Reading, Mass. (1968).

[K] KATO T. : Perturbation Theory for Linear Operators, 2nd edn., Springer, New York (1976).

[R] ROYDEN H.L. : Real Analysis, Macmillan, New York (1963).

[RN] RIESZ F. and SZ.-NAGY B. : Functional Analysis (English translation), Frederick Ungar, New York (1965).

[RS] REED M. and SIMON B. : Methods of Modern Mathematical Physics, Vols I-IV, Academic Press, New York (1972, 1975, 1979 and 1978).

[S] SOBOLEV S.L. : Applications of Functional Analysis in Mathematical Physics, AMS Transl. of Mathem. Monographs, Vol. 7, Providence, R.I. (1963).

[SH] SCHECHTER M. : Spectra of Partial Differential Operators, North-Holland, Amsterdam (1971).

[SW] STEIN E.M. and WEISS G. : Introduction to Fourier Analysis on Euclidean Spaces, Princeton University Press (1971).

[VW] VELO G. and WIGHTMAN A.S., eds : Rigorous Atomic and Molecular Physics, Plenum, New York (1981).

[Z] ZYGMUND A. : Trigonometric Series, Vol. II, Cambridge University Press (1959).

[1] PEARSON D.B. : Spectral Properties and Asymptotic Evolution in Potential Scattering, in [VW].

[2] SINHA K.B. : Ann. Inst. Henri Poincaré A26, 263 (1977).

[3] PEARSON D.B. : Comm. Math. Phys. 60, 13 (1978).

[4] DEIFT P. and SIMON B. : J. Funct. Anal. 23, 218 (1976).

[5] PEARSON D.B. : Comm. Math. Phys. 40, 125 (1975).

[6] AMREIN W.O. and GEORGESCU V. : Helv. Phys. Acta 47, 517 (1974).

[7] PEARSON D.B. : Helv. Phys. Acta 47, 249 (1974).

[8] BAETEMAN M.L. and CHADAN K. : Ann. Inst. Henri Poincaré A24, 1 (1976).

[9] COMBESCURE M. and GINIBRE J. : Ann. Inst. Henri Poincaré A24, 17 (1976).

[10] AMREIN W.O., PEARSON D.B. and WOLLENBERG M. : Helv. Phys. Acta 53, 335 (1980).

[11] PEARSON D.B. : Helv. Phys. Acta 48, 639 (1975).

[12] MOURRE E. : Comm. Math. Phys. 68, 91 (1979).

[13] PERRY P. : Duke Math. J. 47, 187 (1980).

[14] YAFAEV D.R. : On the proof of Enss of asymptotic completeness in potential scattering, preprint (Leningrad 1979).

[15] DOLLARD J.D. : J. Math. Phys. 5, 729 (1964).

[16] MULHERIN D. and ZINNES I.I. : J. Math. Phys. 11, 1402 (1970).

[17] PRUGOVECKI E. and ZORBAS J. : J. Math. Phys. 14, 1398 (1973).

[18] GEORGESCU V. : Méthodes stationnaires pour des potentiels à longue portée à symétrie sphérique, thèse, Université de Genève (1974).

[19] AMREIN W.O. and GEORGESCU V. : Helv. Phys. Acta 46, 635 (1973).

[20] COMBESCURE M. and GINIBRE J. : J. Funct. Anal. 29, 54 (1978).

[21] IKEBE T. and KATO T. : Arch. Ratl. Mech. Anal. 2, 77 (1962).

[22] JÖRGENS K. : Math. Scand. 15, 5 (1964).

[23] RUELLE D. : Nuovo Cimento 61A, 655 (1969).

[24] WILCOX C. : J. Funct. Anal. 12, 257 (1973).

[25] SCHECHTER M. : Letters Math. Phys. 3, 521 (1979).

[26] COOK J.M. : J. Math. and Phys. 36, 82 (1957).

[27] KATO T. : Comm. Math. Phys. 67, 85 (1979).

[28] ENSS V. : Comm. Math. Phys. 61, 285 (1978).

[29] ENSS V. : Comm. Math. Phys. 65, 151 (1979).

[30] ENSS V. : Ann. Phys. 119, 117 (1979).

[31] SIMON B. : Duke Math. J. 46, 119 (1979).

[32] DAVIES E.B. : Duke Math. J. 47, 171 (1980).

[33] GINIBRE J. : La méthode "dépendante du temps" dans le problème de la complétude asymptotique, Preprint (Orsay 1980).

[34] HÖRMANDER L. : Math. Zeitschr. 146, 69 (1976).

[35] KITADA H. : J. Math. Soc. Japan 30, 603 (1978).

[36] PEARSON D.B. : Helv. Phys. Acta 52, 541 (1979).

[37] AMREIN W.O. and PEARSON D.B. : J. Phys. A12, 1469 (1979) and 13, 1259 (1980).

[38] ENSS V. and SIMON B. : Comm. Math. Phys. 76, 177 (1980); Total Cross Sections in Non-Relativistic Scattering Theory, in [GR].

[39] JAUCH J.M. and SINHA K.B. : Helv. Phys. Acta 45, 580 (1972).

[40] MARTIN A. : Comm. Math. Phys. 69, 89 (1979).

[41] JAUCH J.M., LAVINE R.B. and NEWTON R.G. : Helv. Phys. Acta 45, 220 (1972).

[42] DOLLARD J.D. : J. Math. Phys. 14, 708 (1973).

[43] AMREIN W.O., PEARSON D.B. and SINHA K.B. : Nuovo Cimento 52A, 115 (1979).

NOTATION INDEX

SUBJECT INDEX